D1525860

South-East Asian Social Science Monographs

# Malaysia's Industrialization

# Malaysia's Industrialization
## The Quest for Technology

Anuwar Ali

SINGAPORE
OXFORD UNIVERSITY PRESS
OXFORD NEW YORK
1992

*Oxford University Press, Walton Street, Oxford* OX2 6DP

*Oxford New York Toronto*
*Delhi Bombay Calcutta Madras Karachi*
*Kuala Lumpur Singapore Hong Kong Tokyo*
*Nairobi Dar es Salaam Cape Town*
*Melbourne Auckland Madrid*

*and associated companies in*
*Berlin Ibadan*

*Oxford is a trade mark of Oxford University Press*

*Published in the United States*
*by Oxford University Press, New York*



*British Library Cataloguing in Publication Data*

*Data available*

*Library of Congress Cataloging-in-Publication Data*

*Ali, Anuwar,*
*Malaysia's industrialization: the quest for technology/Anuwar Ali.*
*p. cm.—(South-East Asian social science monographs)*
*Includes bibliographical references and index.*
*ISBN 0-19-588601-1:*
*1. Industry and state—Malaysia. 2. Technology and state—Malaysia. 3. Technology transfer—Malaysia. 4. Economic history—1971–1990. 5. Economic history—1990*
*I. Title. II. Series.*
*HD3616.M33A45 1992*
*338'.064'09595—dc20*
*92-15149*
*CIP*

*Typeset by Indah Photosetting Centre Sdn. Bhd., Malaysia*
*Printed in Singapore by Kyodo Printing Co. (S) Pte. Ltd.*
*Published by Oxford University Press Pte. Ltd.,*
*Unit 221, Ubi Avenue 4, Singapore 1440*

# Preface

THE pace of industrial expansion in Malaysia is expected to be more rapid during the 1990s, and this is particularly critical if the country is to achieve the desired transformation of the economy, from one that is semi-industrialized to one fully industrialized. Such a structural transformation would require more than just a set of industrial policies. These must be complemented with strategies to build up the country's technological capability and human resource potential. Both of these are equally critical, and in the context of planning, these two areas have indeed been given substantial focus especially since the Industrial Master Plan was launched in 1986, followed by the Action Plan on Industrial Technology Development in 1990.

Despite all the good intentions of government policies, including those in the Industrial Master Plan and the Action Plan on Industrial Technology Development, there is still much to be done especially at the implementation level. This book is an attempt to examine and assess the many critical issues that are relevant in the context of both industrial development and technology enhancement in this country. However, it is certainly not my aim to provide all the answers to all the related problems of industrialization and technology development in Malaysia, but merely to highlight the relevant issues in the hope that this will stimulate debate and awareness of the constraints faced by Malaysia in wanting to achieve industrial country status.

In writing this book, I am indebted to many individuals from numerous government agencies, research institutes, universities, and the private sector who have assisted me both directly and indirectly. My regular contact and discussion with these individuals, especially during my involvement in the preparation of the Action Plan on Industrial Technology Development with the Ministry of Science, Technology and the Environment and the Institute of Strategic and International Studies (Malaysia) have been most rewarding. My special appreciation goes to those who were directly involved with the preparation of the Action Plan, especially Dr Noordin Sopiee, Datuk Mohd Noordin Hassan, Dr Hamzah Kassim, Dr Rozali Mohamed Ali, (the late) Puan Maimun Din, Mr Kong How Kooi, Mr K. Thiruchelvam, Mr Abdul Rauf Salim, Mr Chan Yuen Hung, and Ms Helen Nesadurai.

Lastly, I am also grateful to Miss Rahimah Mohd Yusof and Mrs Salmah Kasmani for their dedication in typing the manuscript.

*Bangi, Selangor*
*January 1992*

ANUWAR ALI

# Contents

# Tables

# Figures

# Abbreviations and Acronyms

| | |
|---|---|
| ASEAN | Association of South-East Asian Nations |
| EC | European Community |
| EPZs | Export Processing Zones |
| FTZs | Free Trade Zones |
| GDP | Gross domestic product |
| GNP | Gross national product |
| HICOM | Heavy Industries Corporation of Malaysia |
| HRD | Human resource development |
| IMP | Industrial Master Plan |
| IRPA | Intensification of Research in Priority Areas |
| ITAF | Industrial Technical Assistance Fund |
| LUR | Labour Utilisation Relief |
| MARDI | Malaysian Agricultural Research and Development Institute |
| MIDA | Malaysian Industrial Development Authority |
| MIMOS | Malaysian Institute of Microelectronic Systems |
| MNCs | Multinational corporations |
| MOSTE | Ministry of Science, Technology and the Environment |
| NCSRD | National Council for Scientific Research and Development |
| NEP | New Economic Policy |
| NFPEs | Non-financial public enterprises |
| NICs | Newly Industrialized Countries |
| NTBFs | New technology-based firms |
| PORIM | Palm Oil Research Institute of Malaysia |
| R & D | Research and development |
| RBIs | Resource-based industries |
| RRIM | Rubber Research Institute of Malaysia |
| S & T | Science and technology |
| SIRIM | Standards and Industrial Research Institute of Malaysia |
| SMIs | Small- and medium-scale industries |
| TTU | Technology Transfer Unit |
| UNCTAD | United Nations Conference on Trade and Development |
| UNCTC | United Nations Centre on Transnational Corporations |

# 1
# Introduction

THE relatively high growth rates experienced by the Malaysian economy in general and the manufacturing sector in particular during the 1970s gave rise to a lot of hope and expectations regarding the sector's capacity to trigger the economy into a phase of sustainable growth. This was further substantiated by the success of newly industrialized countries like South Korea and Taiwan during the same decade. They became models for export-led industrialization for other developing countries as the latter invariably tried to replicate their successful policies or strategies. However, these expectations were short-lived when, with the coming of the 1980s, the impact of the world economic recession, especially during the 1985–6 period, was extensively felt in Malaysia giving rise to lower growth rates in almost all sectors of the economy including manufacturing. Despite this adverse experience, economic planners still place a lot of hope on the manufacturing sector to boost overall economic growth.

The launching of the Industrial Master Plan (IMP) in 1986 was a reflection of the Malaysian government's concern at that time regarding the state of industrial development, and more importantly, the future role of the manufacturing sector. The sector was expected to become a major catalyst not only to enhance employment opportunities within the country but also to spearhead the country's drive towards industrialized nation status by the end of the twentieth century. It was in this sense, as shall be seen in the following chapters, that so much focus was given to export-led growth and human resource development.

In facing this challenge, the IMP rightly identified a number of constraints which have to be overcome by the manufacturing sector through judicious state planning and appropriate intervention while creating a conducive environment for greater private sector participation. It is indeed the balance between state planning and intervention on the one hand and private sector initiative on the other that will critically influence the pace and nature of future industrialization. Even this balancing act necessitates an effective policy-making institutional framework within the state. One of the principal constraints relates to the dearth of technological capability within domestic industries and the scarcity of industrial research and development (R & D) activities. Domestic industries, in this respect, refer particularly to Malaysian-owned enterprises involved in manufacturing rather than those owned by foreign firms or multinational corporations. In terms of policy initiative, this has been partly answered by the formulation of the Action Plan for Industrial Technology Development that was launched by the government in 1990. Of course, there is much to be done at the implementation level; this, as indicated earlier, will require an enhanced institutional framework within the state.

This book will principally examine the country's potential capacity to propel itself into a new decade of sustained manufacturing growth through the enhancement of industrial technology and the indigenous development of R & D activities. Its main objective will therefore be focused on the challenges which Malaysia has to face in expanding its industrial base in the light of ever changing global investment and technology trends. This also implies that one has to assess the present weaknesses and constraints; in this process, new strategies, options, and measures will be put forward to strengthen the industrialization process through the enhancement of science and technology.

While most chapters of the book will deal with the critical relationship between future industrialization and the need for progress in science and technology (S & T), Chapter 2 begins by examining the process of industrial growth in the country since Independence in 1957, tracing its expansion in relation to the policy adjustments that were initiated by the government right from the import-substitution phase leading towards export-led growth and a heavy industrialization programme during the early 1980s. Principally the consequence of substantial inflows of direct foreign investment into the manufacturing sector, emanating mainly from the industrial countries, manufacturing has now surpassed the agricultural sector as the leading sector, playing a more dominant role not only in terms of employment generation but also as a foreign exchange earner. It is the sector's role with respect to the latter that has been increasingly emphasized since the early 1970s.

The launching of the IMP in 1986 ushered in a new landmark in the development of the manufacturing sector. Its rising share in terms of the country's gross national product became more apparent with the relative decline of the agricultural sector as clearly demonstrated by the downward movement in the country's terms of trade with the rest of the world. With this background, Chapter 3 examines the principal objectives of the Plan. It is through the implementation of the IMP that Malaysia wishes to strive for industrialized nation status. Through greater emphasis on manufacturing, that is, through the expansion of higher value-added and technology-intensive industrial activities, Malaysia hopes to 'leap-frog' into that status.

While the IMP emphasizes the importance of export-led growth and thus the need for domestic industries to be competitive in the international markets, it also emphasizes the future development of the resource-based industries. The development of the latter follows from the need, as identified by the IMP, for the development of greater inter-industry linkages within the economy. Although the Plan does not forward specific policy proposals with respect to technology development, it does focus on the critical role of technology to ensure the rapid expansion of domestic industries. While the launching of the Action Plan for Industrial Technology Development, to complement the IMP, was indeed timely, it is its effective implementation that will determine whether the country's aspirations can be fully realized.

Being a small economy which is critically dependent upon world trade, Malaysia has always been subjected to global economic changes that are mainly influenced by the major trading and industrial nations such as Japan, the United States, and the European Community. While there have been plenty of calls for increased inter-ASEAN and South–South trade, Malaysia's export performance is still extensively dependent upon these industrial nations. Also significant is the country's dependence on imports of capital equipment and intermediate goods from these countries given the increasing momentum of domestic manufacturing growth. In this context, Chapter 4 will appropriately examine the pace of internationalization and global trends and their impact on the economic development of the country. In the Asia–Pacific region, there is no doubt that Japan is a major player not only with respect to its potentially large domestic market but also in terms of its investments overseas. Attention should also be drawn to the emergence of the Asian newly industrialized countries (NICs)—South Korea, Taiwan, Hong Kong, and Singapore—as major exporters of manufactured goods and sources of investments and technology.

To ensure that there is sustained industrial growth, it is crucial that domestic industries constantly familiarize themselves with these global changes, particularly in terms of the acquisition and diffusion of new technologies. These are important areas which domestic industries must keep abreast with, while nurturing the capability to identify the priority technology areas that could be developed domestically. Of course, this will require substantial inputs from the industry itself, particularly in terms of R & D facilities and upgrading of skills on the factory floor.

All these global changes re-emphasize the need for Malaysia to enhance its industrial technology development with the speed that is consonant with the ever changing patterns of demand world-wide. For the moment, the major players are the multinational corporations whose manufacturing, or rather assembly, activities for export are mainly located in Malaysia's Free Trade Zones (FTZs). In this respect, Chapter 5 examines the major forms of technological change and how these forms are related to the development of the manufacturing sector locally. The chapter further examines the nature of technology transfer and the extent of its acquisition by domestic industries. Given that the issue of technology transfer is closely related to direct foreign investment, the chapter also examines the importation of technologies by country of origin and the type of industries in which they predominate. Since technology transfer does influence the development of indigenous technology, it is also pertinent that the existing policy framework and approval mechanism and the policy on royalty remunerations for technology transfer are examined in the chapter.

While it is important that the state monitors and assesses the extent of technology transfer, it is also critical that technological change be directed and managed effectively to ensure rapid industrial growth that is principally driven by the enhancement of domestic technological capability. It is in this context that the role of the state is particularly crucial. Chapter 6 examines this role and rationalizes its importance not only in terms of directing the development of science and technology in the country but also in creating the environment that would allow the nurturing of indigenous technologies. Such a role is considered vital even in industrial countries where the state intervenes much less directly compared to most developing countries. Given that in the early 1990s the contribution of the private sector to industrial R & D is still at a nascent stage, it is therefore crucial that the state plays an appropriate role in ensuring that both public sector and industry R & D activities are enhanced in a complementary manner. However, the present role of the state in science and technology (S & T) development is relatively weak while the national S & T framework is still uncoordinated such that S & T programmes are either ineffective or badly implemented, particularly in the area of industrial R & D.

Within an increasingly competitive world economic environment, it is critical that policy-makers are able to define the country's priorities in terms of its industrial technology development, or at least provide the appropriate signals to private industry. The identification of such priorities would also require a reappraisal of the present S & T strategies. Chapter 7 highlights the major issues that are pertinent to industrial technology development, covering areas such as human resource development (HRD), the development of domestic technological capability, the importance of striking a balance between imported and indigenous technologies, the capability to acquire, adapt, and innovate, and the creation of a conducive S & T environment.

While all these issues are important, the development of human resources has taken centre stage since the mid-1980s, arising mainly from the conspicuous lack of skilled labour to cater for the increasing pace of industrial development in the country. Within the ambit of HRD itself, there is a multiplicity of issues that have to be examined in depth, for instance, the education system, the role of tertiary institutions of learning, the involvement of the private sector in education, and industrial training on the factory floor. Chapter 8 is an extension of the preceding chapter in that it highlights a number of important strategies that would support the enhancement of industrial technology in domestic industries. As implied earlier, these include strategies relating to education and training, industry-level training and support for R & D activities, the promotion of diffusion and application of technology, public sector support policies, and a S & T institutional framework. While some of these strategies may have been assessed, and even adopted, by some government agencies, the chapter nevertheless aims to focus on the critical need for the state to provide the link between the supply of and demand for technology within the manufacturing industries. Malaysia is certainly in a position to learn from the experiences of other countries in the region which have been relatively successful in these areas, such as South Korea, Taiwan, and Singapore.

With the launching of the Action Plan for Industrial Technology Development, the role of the manufacturing sector in expanding the country's industrial base is expected to be given greater prominence during the 1990s and beyond. It is therefore critical that Malaysia creates an environment that will allow science and technology to flourish much more effectively compared to the 1970s and 1980s. During these two decades the industrial structure was relatively dependent on both direct foreign investment and imported technologies and there was an absence of critical linkages among the various industrial sub-sectors. If the country is to achieve the industrial nation status it seeks, it must, amongst other things, be able to develop a critical mass that will move the economy towards a sound footing in science and technology. Plans may provide the necessary targets and objectives that have to be met, but, more importantly, plans have to be implemented effectively. This calls for a greater sense of leadership and commitment towards technology enhancement, apart from the fact that the institutional framework for S & T policy-making and assessment must be strengthened, and hence have the capacity to implement strategies, and if necessary, to adjust those strategies in response to domestic or external changes.

# 2
# Industrialization: Policy Adjustments and Economic Impact

## Policy Adjustments Affecting Industrial Investments

SINCE the enactment of the Pioneer Industries Ordinance in 1958, which was meant to spearhead the industrialization effort in Malaysia, a number of strategic policy changes as well as policy adjustments have been introduced to reflect the changing needs of the industrial sector. The pace of manufacturing growth has significant repercussions on other economic indicators. Thus any change in policies have directly or indirectly affected the whole or some parts of the economy. These changes and adjustments can be chronologically divided into three major phases (Figure 2.1).

Phase I: The years immediately after Independence (when the Pioneer Industries Ordinance of 1958 was introduced) up to 1968 during which the emphasis was on import-substitution industries that were mainly established to cater for the domestic market.

Phase II: The period after 1968 (when the Investment Incentives Act was introduced) up to 1980 during which export-led industrialization was given emphasis through the introduction of export-related incentives and the establishment of Free Trade Zones (FTZs) in a number of locations. This period also saw the introduction of the Industrial Co-ordination Act of 1975 as an instrument to achieve the New Economic Policy (NEP) objectives with regard to Bumiputera equity participation and employment in the manufacturing sector.

Phase III: The period after 1980 which coincides with the implementation of the Fourth Malaysia Plan (1981–5), the formulation of the Industrial Master Plan (IMP) in 1986, and the introduction of the Promotion of Investments Act of 1986, which was seen as an important policy instrument to attract more direct foreign investment into the manufacturing sector. It is also during this period that, first, emphasis was given to second-round import-substitution industries including heavy industries, and secondly, recognition was given to the critical need for technology enhancement in domestic industries. Thus, in line with

FIGURE 2.1
Industrial Development and Major Policy Initiatives, 1958–1990

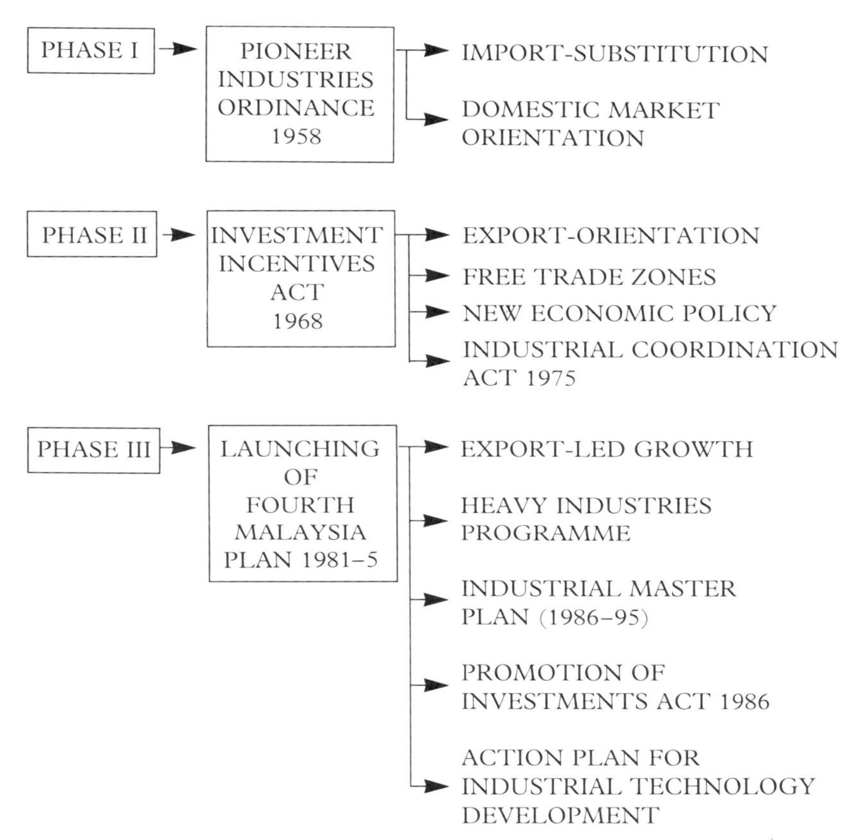


the objectives of the IMP, the Action Plan for Industrial Technology Development was launched in 1990.

A predominant feature of the import-substitution policy during the first phase was the emphasis given to the promotion of industrial development with the private sector, especially foreign investments, playing a pivotal role in the process. The major instruments to promote private investments were the granting of tax holidays, the establishment of industrial estates, the provision of necessary services and infrastructure, and the imposition of tariff protection. In the case of tariff protection, the Customs (Dumping and Subsidies) Ordinance was introduced in 1959 to protect domestic industries against unfair competition; the Ordinance includes the imposition of anti-dumping and countervailing duties. Apart from allowing the nascent industrial base to be broadened, the import-substitution strategy expedited the diversification of the economy, reduced excessive dependence on imported consumer products, utilized some domestic natural resources, and more importantly, as far as the state was concerned, created employment opportunities. Areas where import-substitution had been particularly successful were the renewable resource-based industries such as tobacco, furniture, rubber products, wood products, and food and beverages.

However, the import-substitution phase was at the outset dominated by foreign investments, with little interest on the part of domestic capital to be involved in manufacturing activities. In part, this arose from the pre-Independence period when domestic capital was more inclined to invest in the less risky sectors, mostly the plantation and tin mining sectors. The foreign companies operating during the import-substitution phase were generally producing consumer products with technologies imported from their parent companies. The wholesale importation of technology was indirectly encouraged by the duty exemption given to imported capital equipment and machinery. A similar pattern was duplicated when domestic entrepreneurs or enterprises became involved in manufacturing. In view of their lack of knowledge of technology and inability to master technology, a dependent relationship with their technology suppliers or even their foreign joint-venture partners therefore became apparent. In many ways, this restricts the capability of domestic enterprises to enhance their learning process and technology acquisition.

At the same time, the limited scope for domestic-oriented industries to expand their production capacities had a stifling influence on the economies of scale, which could possibly stimulate investments in technology upgrading within domestic enterprises. In a sense, the import-substitution phase failed to provide the necessary environment for the development of technological capability. The dependence upon imported technologies and expertise had a negative impact on the development of backward and forward linkages, including the development of ancillary and capital goods industries. The incentive regime and the tariff protection given to import-substitution industries inhibited their capacity for technology development. Thus, they lacked the means to increase their competitiveness particularly since tariff protection gave the firms a sense of security within the domain of the domestic market, and hence the disincentive to be more export-oriented (Chan, 1990).

The strategy (also due largely to fiscal incentives that invariably favoured the larger companies), whether intentional or not, had encouraged the more capital-intensive manufacturing projects. This resulted, to a large extent, in manufacturing employment growth lagging behind output growth. Even in terms of employment, the bulk of those employed were principally unskilled or semi-skilled workers. In certain cases, the strategy also attracted less efficient industries that could only survive behind relatively high tariff walls erected on the grounds of their being 'infant industries'. Writing on the growth of the industrial sector during the 1955–67 period, Lo (1972: 95) noted the following:

> The inter-industry relationship was extremely weak as revealed by the structure of sectoral production. Industrial production had been mainly oriented towards import substitution as part and parcel of the strategy of increasing production for domestic use as an engine of growth. Also most of the industries, particularly the so-called pioneer ones, were engaged in simple processing (which consisted

of assembling the imported components and parts rather than real manufacturing) without showing any tendency to advance to higher levels of processing overtime [*sic*]. Thus it is not surprising that in some industries the proportion of import content in the unit cost structure was as high as over 60 per cent. The weaknesses of the prevailing structure of the manufacturing sector could be attributed to a want of coordinated planning in industrial development.

Realizing the limitations of a relatively small domestic market, industrial strategy had to shift from an inward-looking approach to an export-oriented one; this has been particularly apparent since the beginning of the 1970s. Direct foreign investment, as opposed to domestic capital, was accorded an important role not only because it was expected to provide capital, management expertise, and industrial technology but also access to the overseas markets. As such, the impetus for export-led growth has come primarily from the multinational corporations (MNCs) seeking low-cost locations, rather than from domestic enterprises seeking market opportunities. Export expansion that is primarily based upon relatively labour-intensive industries, such as electronics assembly and textiles/garments, largely fails to expand the country's industrial base, thus limiting the capacity for further technology development of the indigenous type. As a consequence of this, the industrial structure which developed became dualistic comprising import-substitution industries and export-oriented industries, without any substantial inter-industry linkages being nurtured.

The second phase involved the production of consumer durables, intermediate inputs, and capital goods. Since most of these products needed to be produced in large quantities utilizing more capital-intensive methods of production, there had to be an adequate supply of highly skilled manpower in order to enjoy economies of scale. Similarly, the second round of export-expansion was a shift from labour-intensive manufactures to more capital- and technology-intensive products. Such products included machinery, motor vehicles, petrochemicals, and other resource-based industries in which the country was expected to develop comparative advantage.

The third phase was a critical one requiring the broadening and deepening of the country's industrial base. When a conscious policy decision was taken to expand the country's more capital-intensive industries during the Fourth Malaysia Plan (1981–5), an important component in this phase was the promotion of heavy industries which started explicitly with the formation of the Heavy Industries Corporation of Malaysia (HICOM) in 1981 (Mohd Saufi Haji Abdullah, 1986: 60–7). Although the development of such industries is still in its embryonic stage, some with high levels of protection, there has been some progress in this direction with the establishment of production facilities in a few basic industries, particularly petrochemicals, fertilizers, automobiles, electronics, cement, and glass. However, the growth of these industries has not been complemented by the expansion of indigenous technological capability. At the same time, the scope for generating ancillary or supporting industrial activities through backward and forward linkages still needs to be fully exploited (Anuwar Ali, 1989b: 298–312). However, as the country moves into a higher phase of its industrialization process, all manufacturing enterprises will without exception require an organizational structure which is capable of eliciting an optimum level of productivity after combining the available factors of production. This is even more crucial in the context of enterprises wishing to venture into heavy or technology-intensive industrial projects. These projects would require not only competent and well-qualified managerial personnel, but also engineering and technical expertise.

In Malaysia, however, the paucity of experienced managerial and technical expertise is still conspicuous, particularly in the manufacturing industries. The output from the existing post-secondary institutions of learning for this category of personnel is inadequate, as well as inappropriate in certain instances, to meet the growing needs of an expanding manufacturing sector. This is partly the reason why the country is slow in nurturing domestic entrepreneurship for the development of small-scale industries. This shortage will have an impact on the industrialization process in many respects. First, it stifles the prospects for enhancing the country's industrial base by limiting the options in terms of going into high-technology and higher value-added industries. Secondly, it limits the capacity to acquire and to adapt imported technologies and thus creates rigidities in terms of innovative capabilities. Thirdly, it lessens the country's potential to compete in international markets for the export of its manufactures.

Various efforts have been initiated to reorientate the education system to give greater emphasis to science and technology or technical education, although the results have been rather mixed. However, the situation is to be reappraised in initiatives taken by the government since the mid-1980s to formulate strategies that favour a technology content. The importance of science and technology has been recognized by the state as an instrument to expand manufacturing activities; this is exemplified in the Fifth Malaysia Plan in which, for the first time, a chapter has been included on the development of science and technology.

With relatively abundant natural resources, it is in the country's interest to utilize these resources fully; within this context the promotion of heavy industries should become an important instrument to increase inter-industry linkages. Such enhancement presupposes the existence of well-developed ancillary or supporting industries, which are still underdeveloped in Malaysia. With increasing industrialization, the demand for components, parts, and intermediate goods would certainly increase, which to a large extent has been satisfied through imports from the industrial countries. In the case of Malaysia's Proton car project, it still has to import a good proportion of its components and parts from Japan, particularly the more complex ones. A situation like this does not assist in the development of ancillary industries to provide the components of the Proton car, although some efforts have been made to establish component manufacturers locally.

Since heavy industries are principally capital-intensive, when seen at the project level this implies opportunity costs as far as labour absorption is concerned. The amount of labour absorbed per unit of capital is certainly less when compared to the less capital-intensive projects. This was indeed a constraint as far as Malaysia was concerned when the rate of unemployment was relatively high during the 1981–5 period. However, these opportunity costs to employment creation are only within a short-term perspective. In the long run, given ample opportunity for such industries to expand their markets, the benefits of heavy industrialization, if they are efficient, must be conceived first in terms of the multiplier effect they will create on the establishment of other industries which will in turn increase employment opportunities. Secondly, they must also be seen in terms of the enhancement of engineering and technical skills among the industrial labour force.

These two elements are extremely critical in determining the success of any industrialization programme. The South Korean experience since the 1970s has been a good barometer for testing these elements. The timing of South Korea's emphasis on heavy industries and export-led growth coincided with a period of booming world trade, during which time there were structural changes within the industrial countries that implicitly forced them to concentrate on higher value-added industries, while leaving the other newly industrialized countries, Singapore, Hong Kong, and Taiwan, to concentrate on the more labour-intensive and mature products. In promoting heavy industries, the South Korean experience showed other countries, such as Malaysia, that its success must be dependent upon a number of prerequisites, including co-ordinated planning at the macro level and detailed planning at the project level. However, even South Korea made mistakes. Commenting on the South Korean experience during the 1970s, Koo and Kwack (1988) wrote:

> A close examination shows that the experiment was in many respects far from successful. Even after a decade of promotion, heavy and chemical industries, with the possible exception of iron and steel and shipbuilding, had failed to achieve efficiency in terms of international competitiveness. Furthermore, they remained largely inefficient and had high cost with unpromising prospects for future exports.

It is in this sense that a developing economy like Malaysia must be able to assess carefully its capacity to initiate major industrial projects such as those associated with heavy industries. As noted by the Industrial Master Plan, a major bottleneck in the development of the heavy engineering industry and the considerable dependence on imports of virtually all components and parts of equipment and machinery arise on account of the lack of basic engineering infrastructure in the country. In fact, a deliberate policy of import protection was instituted, taking the form of quantitative import restrictions on certain capital-intensive products. There is some evidence to indicate that a substantial portion of investment by the non-financial public enterprises (NFPEs) may have been in capital-intensive undertakings and, as indicated earlier, HICOM was clearly established with a mandate to develop capital-intensive projects within the manufacturing sector. The increase in capital-intensity in manufacturing closely parallels the surge in NFPE investments. The objectives were apparently both to diversify the industrial base and to foster infant industries behind protective barriers.

The increases in capital–labour ratios within almost every industry since the early 1980s, combined with rapid increases in investments in the capital-intensive industries, have resulted in the dramatic rise in manufacturing capital-intensity. As labour costs increase while capital costs are subsidized, primarily in the form of tax allowances related to capital, an incentive is thus available for cost-conscious investors to choose relatively more capital-intensive techniques. As illustrated in Table 2.1, fixed assets (as measured at constant prices) per employee have increased substantially within most industries since 1981. The capital–labour ratio for steel, glass products, and non-ferrous metals increased by more than 200 per cent during the 1981–6 period. On the

TABLE 2.1
Increase in Capital-intensity per Employee, 1981–1986

| *Industry* | *Per Cent* |
|---|---|
| Steel | 325.3 |
| Glass products | 265.1 |
| Petroleum refining | 222.5 |
| Non-ferrous metals | 213.6 |
| Transport equipment | 185.9 |
| Non-metallic products | 179.7 |
| Furniture | 139.1 |
| Non-electrical machinery | 136.0 |
| Tobacco products | 126.8 |
| Paper products | 107.9 |
| Metallic products | 92.8 |
| Industrial chemicals | 85.3 |
| Petroleum products | 76.1 |
| Electrical machinery | 71.5 |
| Plastic products | 58.5 |
| Miscellaneous chemicals | 53.0 |
| Printing | 36.7 |
| Beverages | 34.2 |
| Food | 34.1 |
| Wood products | 27.1 |
| Rubber products | 21.5 |
| Textiles | 18.4 |
| Apparel | 10.2 |
| Pottery | –11.9 |
| Scientific instruments | –17.2 |

*Source*: Malaysia (1988a).

other hand, small declines are evident in the scientific apparatus and pottery industries, and the smallest increases occurred in the textiles and apparel sub-sectors. The increase in capital-intensity means that either tremendous domestic savings or growing dependence on direct foreign investment inflows are necessary to generate even small numbers of additional manufacturing jobs. Investment and production costs in heavy industries in developing countries are significantly higher than those of comparable projects in the industrial countries, chiefly because of the relatively small domestic market and thus the absence of economies of scale. In a country such as Malaysia, where the industrial base is still small, this becomes particularly obvious, especially at the initial stages of the heavy industry programme.

High costs are principally due to factors related to the importation of technologies as embodied in the procurement of capital equipment and machinery and the acquisition of technical know-how on production techniques and processes and product development. There are, of course, other factors, including the procurement costs of components, parts, and intermediate inputs and the provision of physical infrastructure and acquiring the appropriate manpower. This last factor would include the hiring of foreign personnel either from the technology suppliers or the foreign joint-venture partners. Most of these costs, at least the first three, are to a large extent beyond the control of the local partners or enterprises. The near monopolistic position of the technology supplier generally reduces the bargaining power of the local partners and this could often lead to higher capital outlays and operating costs that the latter have to bear.

As a matter of policy, it may also be desirable to phase out the import protection which has been introduced to foster the capital-intensive industries. Such protection may be warranted temporarily for infant industries, but these industries may never become efficient and competitive if they are continuously protected; and such protection is certainly costly to consumers and to other end-user industries. It is therefore crucial that the heavy industry programme becomes an integral part of the industrialization process and establishes stronger links with the domestic economy. This means that the limited base of existing heavy industries needs to be expanded, and this requires that vertical linkages with the domestic suppliers/consumers of inputs/outputs of the heavy industry sector be established, and that new industries or expansion programmes of existing projects to meet domestic and overseas demand also be established.

## Manufacturing as a Leading Sector

The expansion of the Malaysian economy in the early 1960s began with agricultural diversification, simultaneously complemented by an industrial strategy which basically promoted the growth of import-substitution industries following the enactment of the Pioneer Industries Ordinance of 1958. Diversification, according to the Second Malaysia Plan

(1971–5), was pursued not only in agriculture but also in other sectors, particularly industry, so that 'the economy was strengthened by a rapid growth of manufacturing production during the decade' (Malaysia, 1971: 15).

As indicated in Table 2.2, the average annual growth rate of the manufacturing sector increased by 11.1 per cent during the 1961–5 period and by 9.9 per cent during the 1966–70 period, with the consequence that the share of manufacturing in gross domestic product (GDP) increased from 8.7 per cent in 1960 to 10.4 per cent in 1965. The growth rate for the sector was not only sustained but increased further during the two succeeding periods, that is, by 11.6 per cent during the 1971–5 period and 13.5 per cent during the 1976–80 period, which were well above the growth rates of GDP during the corresponding periods. (In contrast, the agricultural sector, which had provided the lead since Independence, was witnessing its declining role in its contribution towards the country's output.) By 1975 the share of manufacturing in GDP increased to 16.4 per cent, and in 1980 it was 20.5 per cent.

The promotion of industries during the 1960s was basically meant to increase labour absorption within the economy since employment creation was perceived as a particularly important source of income generation. In the traditional agricultural sector the need for reasonable growth in income puts an upper limit on labour absorption, apart from the structural problems inherent in agriculture. Sufficient employment creation outside the agricultural sector thus assumed even greater significance when, as in Malaysia, the labour force was not only increasing rapidly but was certainly more educated such that its expectations of urban jobs became greater. Although there was considerable uncertainty about the growth rates of employment and labour force during the 1960s, there was a general consensus that employment did not keep pace with the growth of the labour force. Consequently, the open unemployment rate in Peninsular Malaysia increased from about 6 per cent in 1960 to 8 per cent in 1970. This increase suggests that the 6 per cent annual rate of gross national product (GNP) growth recorded in the 1960s was insufficient, even with heavy emphasis on agriculture, to generate an adequate number of jobs (Young et al., 1980: 51).

However, in view of the increasing unemployment problem during the first half of the 1960s, the 1958 Ordinance was repealed and substituted with the Investment Incentives Act of 1968 which was to become the cornerstone of the country's thrust towards industrialization until the early 1980s. Like its predecessor, the 1968 Act was principally meant to encourage more direct foreign investment into the manufacturing sector. There is nothing new in this, since foreign investments have been playing a dominant role in the country's modern sector, in terms of both capital and expertise. In fact, in rationalizing the role of the foreign sector, as early as 1952, Smith (1952: 108) wrote:

> Foreign investors must continue to play an important role in development by establishing new industries and providing representatives to operate and manage

TABLE 2.2

Gross Domestic Product by Sector of Origin, 1960–1980 (percentage)

| | *Share of GDP* | | | | | *Average Annual Growth Rate* | | | |
|---|---|---|---|---|---|---|---|---|---|
| *Sector* | *1960*[ab] | *1965*[ab] | *1970*[b] | *1975*[b] | *1980*[b] | *1961–5* | *1966–70* | *1971–5* | *1976–80* |
| Agriculture, forestry, and fishing | 37.9 | 31.5 | 30.8 | 22.7 | 22.2 | 4.0 | 6.8 | 4.8 | 3.9 |
| Mining and quarrying | 5.9 | 9.0 | 6.3 | 4.6 | 4.6 | 4.5 | 1.1 | 0.4 | 8.9 |
| Manufacturing | 8.7 | 10.4 | 13.4 | 16.4 | 20.5 | 11.1 | 9.9 | 11.6 | 13.5 |
| Construction | 3.0 | 4.1 | 3.8 | 3.8 | 4.5 | 17.9 | 4.1 | 6.6 | 9.8 |
| Electricity, gas, and water | 1.3 | 2.3 | 1.9 | 2.1 | 2.3 | 11.9 | 8.1 | 9.8 | 10.2 |
| Transport and communications | 3.3 | 4.3 | 4.7 | 6.2 | 6.5 | 5.5 | 3.0 | 13.0 | 9.6 |
| Wholesale and retail trade | 15.7 | 15.3 | 13.3 | 12.8 | 12.6 | 6.1 | 3.2 | 6.3 | 8.2 |
| Finance, insurance, and real estate | 6.1 | 6.0 | 8.4 | 8.5 | 8.2 | 6.6 | 6.0 | 7.2 | 8.0 |
| Government services | 6.4 | 5.2 | 11.1 | 12.7 | 13.0 | 4.6 | 5.2 | 10.1 | 9.0 |
| Other services | 11.4 | 10.8 | 2.5 | 2.8 | 2.5 | 7.4 | 4.7 | 9.3 | 6.6 |
| Gross domestic product | – | – | – | – | – | 6.3 | 5.5 | 7.1 | 8.6 |

*Sources*: Malaysia (1965, 1971, and 1981a).

[a]1960 and 1965 figures apply to Peninsular Malaysia only.

[b]The percentage shares of GDP do not add up to 100 per cent because imputed bank service charges and import duties are not considered in the computation.

them. In fact the introduction of Western production techniques and methods of organisation are probably a greater contribution to the future prosperity of Malaya than the actual provision of capital.

Significant structural changes in the composition of output have occurred among and within sectors with the increasing industrial emphasis since the early 1970s. As illustrated in Table 2.3, the manufacturing sector became the largest sector in 1984, marginally superseding the contribution from agriculture, which had traditionally been the largest sector (contributing 20.3 per cent and 20.1 per cent respectively to GDP in that year). In 1985 and 1986, however, manufacturing contribution to total output was lower than agriculture on account of a significant output decline in industries such as electronics, iron and steel, non-metallic mineral products, and petroleum products. As a consequence, the manufacturing sector's contribution to GDP was 19.7 per cent in 1985 while agriculture contributed 20.8 per cent, and their respective contributions in 1986 were 20.9 per cent and 21.4 per cent.

However, after the 1985–6 economic recession, the manufacturing sector's contribution to GDP again exceeded that of agriculture, such that its share in 1987 was 22.5 per cent while agriculture's share was 21.8 per cent. The relatively buoyant economic growth for 1988 increased the manufacturing sector's contribution to GDP to a higher level at 24.4 per cent, while for 1989 its share was 25.6 per cent. In 1990, the percentage share increased further to 27.0 per cent. The sector's increasing importance to domestic economic expansion is clearly reflected in the annual growth rate of the sector, particularly after the economic slow-down of the 1985–6 period. As indicated in Table 2.4, the annual growth rates of the sector since 1987 have far exceeded the growth rates of the other sectors, with 1988 witnessing the highest growth for the sector at 17.6 per cent. During the Sixth Malaysia Plan period, the sector is expected to grow rapidly at an average rate of 11.5 per cent per annum, so that by 1995 its share of total GDP will increase to 32.4 per cent (Malaysia, 1991b: 19).

However, despite the changes in the sectoral composition, the agricultural sector is still crucial as far as the domestic economy is concerned, particularly in terms of employment absorption. Thus, according to the Mid-Term Review of the Fifth Malaysia Plan (Malaysia, 1989a: 131):

> Although the process of structural change had lowered the share of agriculture in gross domestic output due to expansion in the secondary and tertiary components of the GDP, its contribution to economic growth remained positive. Except for 1986 when prices of palm oil fell steeply, earnings of the agriculture sector recovered strongly throughout the Mid-Term Review period. Even in the difficult year of 1986, rubber, pepper and timber prices were already on the rise and the further strengthening of prices in 1987 and 1988 saw rising agricultural terms of trade and income improvements both by smallholders and the corporate agro-based sector.

In 1990, for instance, 1.8 million people were employed in agriculture out of the total working population of 6.6 million. Although its share in

TABLE 2.3

Share in Gross Domestic Product by Industrial Origin, 1983–1990 (percentage)

| *Sector* | *1983* | *1984* | *1985* | *1986* | *1987* | *1988* | *1989* | *1990* |
|---|---|---|---|---|---|---|---|---|
| Agriculture | 21.1 | 20.1 | 20.8 | 21.4 | 21.8 | 21.1 | 20.2 | 18.7 |
| Mining and quarrying | 10.0 | 10.5 | 10.5 | 11.1 | 10.6 | 10.4 | 10.3 | 9.7 |
| Manufacturing | 19.5 | 20.3 | 19.7 | 20.9 | 22.5 | 24.4 | 25.6 | 27.0 |
| Construction | 5.4 | 5.2 | 4.8 | 4.1 | 3.4 | 3.2 | 3.2 | 3.5 |
| Electricity, gas, and water | 1.5 | 1.5 | 1.6 | 1.8 | 1.8 | 1.8 | 1.9 | 1.9 |
| Transport, storage, and communications | 5.8 | 6.0 | 6.4 | 6.6 | 6.7 | 6.7 | 6.7 | 6.9 |
| Wholesale and retail trade | 12.3 | 12.3 | 12.1 | 10.6 | 10.5 | 10.5 | 10.6 | 11.0 |
| Finance, insurance, etc. | 8.5 | 8.5 | 8.9 | 8.8 | 8.9 | 8.9 | 9.1 | 9.7 |
| Government services | 11.8 | 11.8 | 12.2 | 12.5 | 12.4 | 11.8 | 11.4 | 10.7 |
| Other services | 2.2 | 2.1 | 2.3 | 2.3 | 2.3 | 2.2 | 2.1 | 2.1 |
| Less: imputed bank service charges | 2.6 | 2.8 | 3.2 | 3.3 | 3.7 | 4.3 | 4.7 | 5.1 |
| Add: import duties | 3.9 | 4.4 | 3.9 | 3.0 | 2.7 | 3.2 | 3.6 | 3.8 |
| GDP at market price | 100 | 100 | 100 | 100 | 100 | 100 | 100 | 100 |

*Sources*: Malaysia (1989b, 1991b).

TABLE 2.4
Annual Growth Rate of Gross Domestic Product by Sector, 1983–1990 (percentage)

| | *Annual Growth Rate* | | | | | | | |
|---|---|---|---|---|---|---|---|---|
| *Sector* | *1983* | *1984* | *1985* | *1986* | *1987* | *1988* | *1989*[a] | *1990*[a] |
| Agriculture, forestry, livestock, and fishing | −0.6 | 2.8 | 2.5 | 4.0 | 7.4 | 5.2 | 3.1 | 2.6 |
| Mining and quarrying | 15.7 | 13.7 | −1.4 | 7.5 | 0.1 | 6.6 | 6.9 | 2.1 |
| Manufacturing | 7.9 | 12.3 | −3.8 | 7.5 | 13.4 | 17.6 | 13.0 | 10.5 |
| Construction | 10.4 | 4.2 | −8.4 | −14.0 | −11.8 | 2.7 | 8.5 | 12.5 |
| Electricity, gas, and water | 10.7 | 11.5 | 6.5 | 8.3 | 8.0 | 9.2 | 9.2 | 9.0 |
| Transport, storage, and communications | 5.2 | 10.4 | 4.8 | 6.1 | 5.3 | 8.8 | 7.6 | 6.4 |
| Wholesale and retail trade, hotels and restaurants | 7.8 | 8.0 | −2.8 | −11.1 | 4.5 | 8.8 | 8.0 | 7.5 |
| Finance, insurance, real estate, and business services | 8.0 | 7.0 | 4.1 | −0.4 | 6.8 | 9.0 | 9.5 | 7.5 |
| Government services | 5.0 | 7.7 | 2.1 | 4.3 | 4.0 | 3.7 | 4.0 | 3.7 |
| Other services | 4.6 | 4.7 | 4.1 | 4.0 | 3.6 | 3.9 | 4.4 | 4.4 |
| Gross domestic product | 6.3 | 7.8 | −1.0 | 1.2 | 5.3 | 8.7 | 7.6 | 6.5 |

*Source*: Malaysia (1989b).
[a]Figures for 1989 and 1990 are estimates.

total employment had declined over the previous two decades, the percentage share of 27.8 per cent in 1990 was still the largest, followed by the manufacturing sector with 19.5 per cent. However, during the 1986–90 period, the agricultural sector accounted for 7.8 per cent of new employment generated, that is, about 78,000 new jobs out of the total 996,000 jobs created during the period. This indicates an average annual growth of only 0.9 per cent, which was much lower than the annual growth of 3.3 per cent for the whole economy. In contrast, the manufacturing sector accounted for 43.6 per cent of new jobs created (that is, about 434,800 jobs), equivalent to an annual growth of 8.6 per cent (Malaysia, 1991b: 28).

## Increasing Capital-intensity and Employment

During the 1960s manufacturing growth was mainly perceived in terms of its contribution to national output as well as employment generation. However, the relatively more labour-intensive industries did not grow as fast as the relatively capital-intensive ones. The principal reason for this was related to the Pioneer Industries Ordinance of 1958, whose tax exemptions corresponded with the level of capital expenditure. This means that the relatively capital-intensive industries such as petroleum and coal products, beverages, chemical products, non-metallic mineral products, food processing, and the processing of estate-type agricultural products were more likely to benefit from the incentives offered than the relatively labour-intensive industries such as textiles, footwear, furniture, leather products, and wood and paper products.

The capacity of the manufacturing sector to absorb labour was found to be comparatively low. For instance, during the 1962–70 period, while manufacturing value-added increased by 12.0 per cent, manufacturing employment increased by only 4.4 per cent (Malaysia, 1971: 148). This appeared to be unsatisfactory, given the important role that the state had placed on the sector in reducing the rate of unemployment, which stood at a high of 8.0 per cent in 1970.

The capital-intensity of the manufacturing sector, which is often identified with dependence on foreign capital and technologies, seems to indicate that employment growth in the sector will not keep pace with the increasing supply of labour in the market-place. Although foreign firms may be associated with better management techniques, they are not keen to employ large numbers of workers for a number of reasons. Lim (1975: 213–24), for instance, argues that since their wages are expected to be higher than those paid by locally controlled firms, this has affected the hiring of labour itself. The relatively higher wages paid by foreign firms may have a demonstration effect on the employees of locally owned firms who may consequently be discouraged from installing more labour-intensive techniques. However, this argument cannot be regarded as conclusive. In the smaller manufacturing firms, which are generally Malaysian-owned, there is evidence to indicate that 'wage rates have tended to remain stagnant and fall behind other firms due to the absence of any pressure to raise wages from organised labour' (Arudsothy, 1977: 89–96).

As indicated earlier in the chapter, over the years the manufacturing sector has experienced an increasing capital-intensity as demonstrated by the increasing value-added per employee in almost all the industrial sub-sectors, particularly in medium- and large-scale enterprises, but less so within small-scale enterprises (Anuwar Ali, 1989a). Since the latter are generally labour-intensive and less skill-intensive than their larger counterparts, it implies that there exists a 'dualistic technological structure' within the manufacturing sector. This becomes more pronounced when comparisons are made between domestic manufacturing enterprises and the larger foreign-owned enterprises. More significantly, in terms of industrial growth and technology development, the substantial increase in the value-added per employee for the large-scale enterprises is related to three important factors. First, the size of manufacturing enterprises as a whole has generally increased, partly because of the nature of fiscal incentives which favour such a development. Secondly, there is a change in industry composition favouring those with high value-added per employee or capital-intensive industries. Thirdly, large manufacturing enterprises enjoy technological and management economies of scale. Added to these three factors is the absence of any meaningful attempt to enhance the role of the small-scale industries, particularly in the 1960s and 1970s. It was only during the Fifth Malaysia Plan (1986–90) that there was a clearer perspective on the role of this sector.

It has also been observed that the increase in non-wage value-added per employee for all enterprise sizes is significantly higher than the increase in wage value-added per employee, thus implying that the increase in physical capital is more prevalent than the increase in human capital, and reflecting an employment structure within the manufacturing sector which exhibits a high proportion of semi-skilled and unskilled workers. The larger firms tend to resort to 'mechanization' of their production processes given that the supply of highly skilled labour is relatively limited. Despite the problem of low employment absorption, this situation, to a large extent, has been responsible for the upward trends in average labour productivity in manufacturing during the 1970s and 1980s.

The increase in such productivity can also be associated with the expansion of existing large firms, either through increased production volumes or the inclusion of new product lines, and the entry of new firms with larger capital outlays and capacities which have the propensity to introduce capital equipment that embodies state of the art technology into the country. This, unfortunately, produces a negative effect that results in the squeezing out of some of the smaller firms which become less competitive. In the process of achieving rapid industrial expansion, this becomes unavoidable since the adoption of relatively capital-intensive and more efficient techniques originating from the industrial countries has given rise to this 'backwash' effect.

However, for many industries, the adoption of capital-intensive techniques is essential to achieve economies of scale in order to get ahead of their competitors in both domestic and international markets. With technical progress and scale economies, the efficiency of modern production techniques is bound to increase in comparison with the earlier vintage techniques. The relatively low cost of labour in relation to capital is generally inadequate to compensate for the lower efficiency of the older techniques. The superiority of modern technology is enhanced if optimum use of inputs and product quality and acceptability are taken into account. A significant implication in terms of future industrial strategy is the need to have a strong science and technology infrastructure so as to enhance the domestic capacity to acquire and adapt imported technologies to suit domestic needs.

Realizing the seriousness of the unemployment problem in the late 1960s, the government replaced the Pioneer Industries Ordinance with the Investment Incentives Act in 1968 under which additional incentives were granted, including an investment tax credit, accelerated depreciation allowance, and export incentives. In addition, tariff protection and exemption from import duties and surtax were granted to facilitate the establishment of new manufacturing enterprises. These provisions signalled a strategic switch in emphasis from import-substitution to export-oriented industrialization. In this case, the role of direct foreign investment is further emphasized, and even the role of the multinational corporations has been justified on the grounds of employment absorption. Thus, according to Jesudason (1989: 167–8):

> The state also wanted multinationals to provide high levels of employment. The leaders and policy-makers in Malaysia have been sensitive to the undesirable consequences of high unemployment levels in society. As an electoral regime, the pressure of mass demands could not be so easily dismissed politically as in more authoritarian societies. The second prong of the NEP promised to eradicate poverty irrespective of race. The main strategy to appease the poorer sectors of the population was to increase the rate of employment creation, particularly in the envisaged growth sector, manufacturing.

Although the newly introduced Investment Incentives Act had a wider range of incentives, most of these incentives were, as in previous Acts, based fundamentally on the level of capital expenditure. However, in July 1971 an amendment was introduced to include tax exemption based on the number of workers employed. The Labour Utilisation Relief (LUR) was intended to expand employment opportunities within the sector while simultaneously encouraging the establishment of export-based activities. The introduction of the LUR coincided with the dropping of the payroll tax of 2 per cent on the grounds that its abolition would induce manufacturing enterprises to utilize more labour. The LUR was the only employment incentive scheme of its kind formulated and implemented during the 1970s. However, its net impact on employment creation was doubtful. It was provided along with other incentives such as pioneer status and the investment tax incentive. A firm which tended to be capital-intensive and which expected reasonable profits soon after production would be likely to choose the pioneer status incentives. A highly capital-intensive firm with a lengthy gestation period would prefer the investment tax incentive, while a firm that opted for the LUR was likely to be inherently labour-intensive in the first instance. A good example would be the textile industry.

However, in spite of its small base, the ability of the manufacturing sector during the 1970s to generate employment opportunities was relatively impressive, as the growth rate of employment in the sector surpassed the overall employment growth rate. As indicated in Table 2.5, the average annual growth rate of manufacturing employment during the 1970s was almost double that of total employment growth, that is, 8.2 per cent during the 1971–5 period and 7.0 per cent during the 1976–80 period compared to 4.6 per cent and 3.7 per cent respectively for total employment growth. In comparison, the average annual growth rates for the agricultural sector were 2.3 per cent and 1.5 per cent for the two periods.

In absolute terms, the manufacturing sector created 454,100 new jobs (or slightly over 45,000 jobs annually), representing 30.7 per cent of the total new employment created during the period. Consequently, in terms of overall employment, the share of the manufacturing sector in total employment increased from 11.4 per cent in 1970 to 15.8 per cent in 1980, while that of agriculture declined from 50.5 per cent in 1970 to 40.6 per cent in 1980. The textile, electrical, rubber and plastic products, food processing, and wood products industries together accounted for the bulk of the total number of new jobs created.

The relatively rapid employment absorption in the manufacturing sector during the 1970s could not, however, be sustained during the 1980s. The economic recession of the early 1980s slowed down the rate of employment absorption markedly. The annual growth of manufacturing employment was only 2.5 per cent during 1981–5 compared to 3.1 per cent for the whole economy. About 100,300 new jobs were created, representing only 12.4 per cent of the total new employment generated. The main industries responsible for most of the new employment were wearing apparel, paper products, petroleum products, pottery, non-ferrous metal products, and electrical machinery. Industries such as textiles, wood products, furniture, food processing, chemicals, rubber products, and leather and footwear were especially affected by the economic slow-down and experienced absolute declines in employment. The global economic recession seemed to have a profound impact on the export-oriented firms, particularly in industries such as electronics, textiles, and plywood and wood-based products. They were forced to retrench some of their workers. The retrenchments which took place from the middle of 1985 indirectly increased the overall unemployment rate from 5.0 per cent in 1981 to 6.9 per cent in 1985. In

TABLE 2.5
Employment and Employment Growth by Sector, 1965–1980 (percentage)

| Sector | Share of Total Employment | | | | Average Annual Growth Rate | | |
|---|---|---|---|---|---|---|---|
| | *1965*[c] | *1970* | *1975* | *1980* | *1965–70* | *1971–5* | *1976–80* |
| Agriculture, forestry, and fishing | 52.1 | 50.5 | 45.3 | 40.6 | 1.5 | 2.3 | 1.5 |
| Mining and quarrying | 2.5 | 2.6 | 2.1 | 1.7 | −0.6 | −0.1 | 0.3 |
| Manufacturing | 8.4 | 11.4 | 13.5 | 15.8 | 4.5 | 8.2 | 7.0 |
| Construction | 3.5 | 4.0 | 4.4 | 5.2 | 2.7 | 6.6 | 7.0 |
| Electricity, water, and gas | 0.6 | 0.8 | 0.8 | 1.0 | 3.5 | 4.6 | 8.3 |
| Transport and communications | 3.9 | 3.4 | 3.9 | 3.8 | 1.7 | 7.5 | 3.1 |
| Commerce[a] | 11.1 | 11.8 | 12.7 | 13.7 | 3.4 | 7.0 | 5.6 |
| Services[b] | 17.9 | 15.5 | 17.3 | 18.2 | 4.5 | 7.9 | 5.3 |
| Total employment | 100.0 | 100.0 | 100.0 | 100.0 | 2.6 | 4.6 | 3.7 |

*Sources*: Malaysia (1971, 1981a).
[a]Commerce includes wholesale and retail trade, hotels and restaurants, finance, real estate, and business services.
[b]Services include government and other services.
[c]1965 figures apply to Peninsular Malaysia only.

the following year, out of the total 20,212 employees retrenched within the economy, 54 per cent of them were from the manufacturing sector (Malaysia, 1988b: 20).

The strengthening of the economy since late 1986 has, however, improved the overall employment prospects within the economy. As indicated in Table 2.6, total employment was estimated to have increased by 3.3 per cent during the 1986–90 period, which was higher than the growth rate of the labour force at 3.1 per cent. While the annual growth rate of 8.6 per cent for the manufacturing sector was the highest during the 1986–90 period, both the mining and quarrying as well as the construction sectors experienced negative rates of employment absorption during the period, arising mainly from the impact of the 1985–6 economic recession.

The manufacturing sector accounted for 43.6 per cent of the total employment increase, amounting to 434,800 new jobs, during the 1986–90 period. This expansion in manufacturing employment was mainly attributed to production increases in industries such as rubber-based products, wood-based products, textiles, electrical and electronic products, plastic products, transport equipment, and manufacture of machinery. During the period, each of these industry groups registered an average growth rate in employment of between 14 and 20 per cent annually (Malaysia, 1991b: 131). With these changing trends in sectoral labour absorption, the manufacturing sector experienced an increase in its employment share from 15.2 per cent in 1985 to 19.5 per cent in 1990, and this is expected to increase further to 21.9 per cent in 1995. According to the Second Outline Perspective Plan (1991–2000), the share is expected to increase further to 23.9 per cent by the year 2000. The share of agriculture, however, declined from 31.3 per cent in 1985 to 27.8 per cent in 1990 and this is expected to decline further to 23.5 per cent in 1995. However, the increase in employment opportunities during the 1986–90 period did not have any substantial impact on the level of unemployment. In fact, the unemployment rate declined only marginally from 6.9 per cent in 1985 to 6.0 per cent in 1990, although, with increasing optimism in terms of the economy's growth prospects, the rate is expected to decline to 4.5 per cent in 1995.

An examination of the changing demand for labour in the manufacturing sector during the 1970s and the early 1980s shows a number of interesting features. First, while there was considerable expansion in manufacturing employment, it was mostly concentrated in a few export-oriented activities, such as the electrical machinery (particularly electronics), textile, wearing apparel, and wood-based industries. During the 1973–81 period, for instance, nearly 50 per cent of the 236,211 new jobs created in manufacturing were from these four industry groups, although the industrial strategy had a strong domestic-expansion bias. The electrical machinery industry alone generated 24 per cent of this increase; the average establishment size in these industries was quite large.

TABLE 2.6

Employment and Employment Growth by Sector, 1985–1995

| | Increase ('000) | | | | | Average Annual Growth Rate (%) | |
|---|---|---|---|---|---|---|---|
| *Sector* | *1985* | *1990* | *1995* | *1986–90* | *1991–5* | *1986–90* | *1991–5* |
| Agriculture, forestry, livestock, and fishing | 1,759.6 | 1,837.6 | 1,821.9 | 78.0 | –15.7 | 0.9 | –0.2 |
| Mining and quarrying | 44.4 | 39.1 | 40.7 | –5.3 | 1.6 | –2.5 | 0.8 |
| Manufacturing | 855.4 | 1,290.2 | 1,699.1 | 434.8 | 408.9 | 8.6 | 5.7 |
| Construction | 429.4 | 426.9 | 547.5 | –2.5 | 120.6 | –0.1 | 5.1 |
| Non-government services | 1,716.3 | 2,177.0 | 2,770.9 | 460.7 | 593.9 | 4.9 | 4.9 |
| Government services | 819.5 | 850.2 | 872.2 | 30.7 | 22.0 | 0.7 | 0.5 |
| Total | 5,624.6 | 6,621.0 | 7,752.3 | 996.4 | 1,131.3 | 3.3 | 3.2 |
| Labour force | 6,039.1 | 7,046.5 | 8,114.0 | 1,007.4 | 1,067.5 | 3.1 | 2.9 |
| Unemployment | 414.5 | 425.5 | 361.7 | 11.0 | –63.8 | | |
| Unemployment rate (%) | 6.9 | 6.0 | 4.5 | | | | |

*Source*: Malaysia (1991b).

*Note*: Non-government services include electricity, gas, and water; transport, storage, and communications; wholesale and retail trade; hotels and restaurants; finance, insurance, real estate, and business services; and other services.

Secondly, it was only during the 1982–5 period that export expansion became a very much more significant source of manufacturing growth than import-substitution. Unfortunately, this happened during a period when economic recession set in, with serious adverse effects on employment absorption. For example, between 1983 and 1985 employment in the textile, wood products, and electrical machinery industries decreased by 17.3 per cent, 18.5 per cent, and 6.0 per cent respectively. The principal reason for this was the drop in world demand for these products and the increasing capital-intensity and automation in their production, particularly in the electronics sub-sector.

Thirdly, export-oriented industries were vulnerable to external trade influences and could not avoid the effect of the global recession and the accompanying retrenchment of workers. The first spate of retrenchments occurred during the 1974–5 period. When 2,230 workers within the manufacturing sector were laid off, about 50 per cent of them were electronics workers. In the middle of the 1980s, almost 90 per cent of the workers retrenched were from the electronics, textile, and wood-based industries. Thus, if the focus is given to only a few export-oriented industries, the country's economic well-being could become very vulnerable to externally induced economic slow-downs.

Fourthly, the labour absorption rate of manufacturing in the 1980s was lower than in the 1970s. The main reason was the changing pattern of production in both the established and the new manufacturing firms. Since many firms that were granted pioneer status in the early 1970s had already ended their tax exemption period by the end of the decade or the early 1980s, they started to rationalize their production technologies/processes which resulted in relatively fewer workers being employed. In addition, the liberal policies of the government since 1986 to attract direct foreign investment into industries requiring higher levels of technology meant increasing capital-intensity and labour productivity, thus slowing down the rate of labour absorption.

## The Increase in Manufactured Exports in the 1970s and 1980s

The changing export structure has, during the 1970s and 1980s, been influenced to a large extent by the increasing role of export-oriented industries. During the 1980s, such a change was reflected in the increasing role of manufactures in the country's total exports. Table 2.7 indicates that the share of manufactures in total exports had increased substantially during the 1980s, with a 31.6 per cent share in 1984 increasing to 60.4 per cent in 1990. These increases were made at the expense of the country's primary commodities, particularly exports of crude petroleum, rubber, and palm oil. In fact, during the 1986–90 period, the average annual growth rate of manufactured exports at 31.0 per cent far exceeded the growth rates of other commodities. By 1995 the percentage share of manufactured exports to total exports is projected to reach 75.0 per cent (Malaysia, 1991b: 23).

TABLE 2.7
Major Exports, 1984–1990 ($ million)

| *Exports* | *1984* | *1985* | *1986* | *1987* | *1988* | *1989*[a] | *1990* |
|---|---|---|---|---|---|---|---|
| Rubber | 3,672 | 2,872 | 3,183 | 3,915 | 5,256 | 3,949 | 3,028 |
| | (9.8) | (7.6) | (8.5) | (8.6) | (9.6) | (5.8) | (3.8) |
| Tin | 1,162 | 1,648 | 650 | 839 | 911 | 1,161 | 902 |
| | (3.0) | (4.4) | (1.7) | (1.9) | (1.7) | (1.7) | (1.1) |
| Crude petroleum | 8,737 | 8,698 | 5,401 | 6,290 | 6,128 | 7,857 | 10,637 |
| | (22.7) | (23.1) | (14.4) | (14.1) | (11.2) | (11.6) | (13.4) |
| Palm oil | 4,531 | 3,951 | 3,010 | 3,279 | 4,528 | 4,692 | 4,399 |
| | (11.8) | (10.5) | (8.0) | (7.3) | (8.3) | (6.9) | (5.5) |
| Sawn logs | 2,790 | 2,748 | 2,847 | 4,274 | 4,010 | 4,354 | 4,041 |
| | (7.3) | (7.3) | (7.6) | (9.6) | (7.3) | (6.4) | (5.1) |
| Sawn timber | 994 | 973 | 1,191 | 1,652 | 1,844 | 2,904 | 3,065 |
| | (2.6) | (2.6) | (3.2) | (3.7) | (3.4) | (4.3) | (3.9) |
| Manufactures | 12,164 | 12,111 | 15,329 | 20,216 | 27,085 | 36,661 | 48,047 |
| | (31.6) | (32.2) | (40.8) | (45.2) | (49.6) | (54.1) | (60.4) |

*Sources*: Malaysia (1989b, 1991b).
*Note*: Figures in parentheses indicate percentage of total exports during the relevant year.
[a]1989 figures are estimates obtained from Malaysia (1990a: 201).

On average, the electrical and electronic machinery sub-sector has accounted for just over 50 per cent of manufactured exports since 1985, while, in contrast, its contribution was only 3 per cent in 1970 (Table 2.8). The other major contributors are the textile and footwear sub-sectors, accounting for slightly over 10 per cent of total manufactured exports during the same period. As implied earlier, much of the increase in manufactured exports is traceable to the establishment of the FTZs since the early 1970s and the inflow of direct foreign investment as well as international subcontracting arrangements where labour-intensive processes have been relocated to Malaysia.

During the 1986–90 period, exports of manufactured goods expanded at an average rate of 31.0 per cent per annum and the country's total export value increased from $12,470.8 million in 1985 to $48,047.1 million in 1990 (Malaysia, 1991b: 130). While electrical goods and electronics continued to be the major component, new products, such as rubber products and iron and steel, began to make inroads into the export market. The increasing importance of manufactures in terms of the country's exports must, however, be seen in the context of the country's changing import structure where, with increasing industrialization, imports of machinery and capital equipment have simultaneously increased at a relatively rapid rate during the 1980s. Table 2.9 indicates the substantial increase in the import of investment goods since 1970, accounting for 26.6 per cent of total imports in 1970, increasing to 30.0 per cent in 1980 and to 34.3 per cent in 1989. Machinery accounted for a significant share of this total. Also important was the rapid increase in the import of intermediate goods (especially for the

TABLE 2.8
Exports of Manufactured Goods, 1970–1989

| *Manufacturing Sub-sector* | *1970* | | *1980* | | *1985* | | *1986* | | *1987* | | *1988* | | *1989*[a] | |
|---|---|---|---|---|---|---|---|---|---|---|---|---|---|---|
| | *($ million)* | *(%)* | *($ million)* | *(%)* | *($ million)* | *(%)* | *($ million)* | *(%)* | *($ million)* | *(%)* | *($ million)* | *(%)* | *($ million)* | *(%)* |
| Food, beverages, and tobacco | 112 | 18 | 475 | 8 | 594 | 5 | 753 | 5 | 880 | 4 | 1,043 | 4 | 736 | 4 |
| Textiles and footwear | 40 | 7 | 806 | 13 | 1,289 | 10 | 1,644 | 11 | 2,285 | 11 | 2,958 | 11 | 2,032 | 10 |
| Wood products | 88 | 14 | 467 | 8 | 363 | 3 | 534 | 3 | 849 | 4 | 918 | 3 | 582 | 3 |
| Rubber products | 17 | 3 | 84 | 1 | 113 | 1 | 155 | 1 | 243 | 1 | 326 | 1 | 213 | 1 |
| Chemical and petroleum products | 197 | 32 | 361 | 6 | 1,412 | 12 | 1,239 | 8 | 1,483 | 7 | 1,912 | 7 | 1,261 | 6 |
| Non-metallic mineral products | 20 | 3 | 61 | 1 | 150 | 1 | 191 | 1 | 302 | 1 | 444 | 2 | 372 | 2 |
| Iron and steel, and metal products | 26 | 4 | 161 | 3 | 300 | 2 | 444 | 3 | 694 | 3 | 1,000 | 4 | 782 | 4 |
| Electrical, electronic machinery, and appliances | 17 | 3 | 2,832 | 46 | 6,028 | 50 | 7,976 | 52 | 10,251 | 51 | 14,039 | 52 | 10,164 | 52 |
| Other machinery and transport equipment | 68 | 11 | 407 | 7 | 1,031 | 9 | 1,034 | 7 | 1,447 | 7 | 1,625 | 6 | 1,308 | 7 |
| Others manufactures | 27 | 5 | 447 | 7 | 831 | 7 | 1,359 | 9 | 1,782 | 9 | 2,820 | 10 | 2,197 | 11 |
| Total | 612 | 100 | 6,101 | 100 | 12,111 | 100 | 15,329 | 100 | 20,216 | 100 | 27,085 | 100 | 19,647 | 100 |

*Source*: Malaysia (1989b).
[a]Data for 1989 are preliminary estimates for January–July.

TABLE 2.9

Gross Imports by Economic Function, 1970–1989 ($ million)

| Items | 1970 | | 1975 | | 1980 | | 1988 | | 1989[a] | |
|---|---|---|---|---|---|---|---|---|---|---|
| | *$ million* | *(%)* | *$ million* | *(%)* | *$ million* | *(%)* | *$ million* | *(%)* | *$ million* | *(%)* |
| Consumption goods | 1,212 | 28.0 | 1,720 | 20.0 | 4,325 | 18.4 | 10,246 | 23.6 | 13,359 | 22.0 |
| Food | 490 | 11.3 | 615 | 7.2 | 1,177 | 5.0 | 2,344 | 5.4 | 2,912 | 4.8 |
| Beverage and tobacco | 74 | 1.7 | 85 | 1.0 | – | – | 217 | 0.5 | 250 | 0.4 |
| Consumer durables | 134 | 3.1 | 260 | 3.0 | 992 | 4.2 | 3,386 | 7.8 | 4,617 | 7.6 |
| Others | 514 | 11.9 | 760 | 8.8 | 2,156 | 9.2 | 4,299 | 9.9 | 5,580 | 9.2 |
| Investment goods | 1,152 | 26.6 | 2,740 | 31.9 | 7,030 | 29.9 | 12,677 | 29.2 | 20,890 | 34.3 |
| Machinery | 455 | 10.5 | 950 | 11.1 | 2,578 | 11.0 | 3,864 | 8.9 | 6,400 | 10.5 |
| Transport equipment | 151 | 3.5 | 250 | 2.9 | 919 | 3.9 | 1,433 | 3.3 | 3,734 | 6.1 |
| Metal products | 283 | 6.5 | 550 | 6.4 | 1,767 | 7.5 | 2,865 | 6.6 | 4,021 | 6.6 |
| Others | 263 | 6.1 | 990 | 11.5 | 1,766 | 7.5 | 4,515 | 10.4 | 6,735 | 11.1 |
| Intermediate goods | 1,572 | 36.4 | 3,726 | 43.4 | 11,695 | 49.9 | 20;057 | 46.2 | 25,962 | 42.6 |
| For manufacturing | 893 | 20.7 | 2,017 | 23.5 | 6,670 | 28.4 | 15,151 | 34.9 | 20,031 | 32.9 |
| For construction | 83 | 1.9 | 170 | 2.0 | 580 | 2.5 | 1,129 | 2.6 | 1,545 | 2.5 |
| For agriculture | 158 | 3.6 | 330 | 3.9 | 899 | 3.8 | 1,042 | 2.4 | 1,112 | 1.8 |
| Crude petroleum | 215 | 5.0 | 665 | 7.7 | 1,890 | 8.1 | 410 | 0.9 | 310 | 0.5 |
| Others | 223 | 5.2 | 544 | 6.3 | 1,656 | 7.1 | 2,325 | 5.4 | 2,964 | 4.9 |
| Imports for re-export | 349 | 9.0 | 405 | 4.7 | 406 | 1.7 | 433 | 1.0 | 687 | 1.1 |
| Total | 4,285 | 100.0 | 8,591 | 100.0 | 23,456 | 100.0 | 43,413 | 100.0 | 60,898 | 100.0 |

*Sources*: Bank Negara Malaysia Annual Reports, 1970, 1975, 1980, 1988, and 1989.

[a]Data for 1989 are preliminary estimates.

TABLE 2.10
Technology Imports and Macroeconomic Performance in 53 Developing Countries: Growth Rates (percentage)

| | *Capital Goods Imports* | | | *Gross Domestic Product* | | | *Gross Fixed Capital Formation* | | |
|---|---|---|---|---|---|---|---|---|---|
| | *1970–81* | *1981–6* | *1970–86* | *1970–81* | *1981–6* | *1970–86* | *1970–81* | *1981–6* | *1970–86* |
| Sample of 53 developing countries | 22.6 | –2.1 | 12.8 | 5.6 | 1.8 | 4.3 | 8.0 | –1.6 | 4.7 |
| Of which: | | | | | | | | | |
| By region | | | | | | | | | |
| Developing Africa | 24.8 | –6.9 | 12.2 | 4.4 | 1.6 | 3.5 | 10.5 | –0.6 | 6.7 |
| Developing America | 20.7 | –6.2 | 10.2 | 5.6 | 0.5 | 3.9 | 6.5 | –5.8 | 2.2 |
| Developing Asia | 23.9 | 4.2 | 15.8 | 6.3 | 3.8 | 5.5 | 9.1 | 3.1 | 7.1 |
| By analytical group | | | | | | | | | |
| Major petroleum exporters | 26.1 | –8.5 | 12.4 | 6.5 | 0.3 | 4.4 | 11.1 | –4.3 | 5.7 |
| Major exporters of manufactures | 21.7 | 4.1 | 14.5 | 5.8 | 2.0 | 4.5 | 6.0 | –1.0 | 3.6 |
| Least developed countries | 20.9 | –2.0 | 11.9 | 1.2 | 1.9 | 1.4 | 4.7 | 4.3 | 4.6 |
| Remaining countries | 20.0 | –0.2 | 12.0 | 5.1 | 2.9 | 4.3 | 7.0 | 0.5 | 4.8 |
| By income group | | | | | | | | | |
| Over US$1,500 | 23.1 | –1.2 | 13.4 | 6.1 | 1.4 | 4.5 | 7.6 | –2.5 | 4.1 |
| Between US$500 and US$1,500 | 22.0 | –8.9 | 10.0 | 5.6 | 0.3 | 3.8 | 9.9 | –4.9 | 4.7 |
| Below US$500 | 22.0 | 2.6 | 14.2 | 4.7 | 3.7 | 4.4 | 7.6 | 3.4 | 6.2 |

*Source*: UNCTAD (1988).

manufacturing sector), increasing its share from 36.4 per cent of total imports in 1970 to 42.6 per cent in 1989. This reflects a situation in which inter-industry linkages within domestic industries are still underdeveloped.

Since almost all of these imports are from the industrial countries (Japan, the biggest exporter of capital equipment to Malaysia, accounting for about 35 per cent of the total), the growth of domestic industries is continually interlocked with industrial country technology (Stewart, 1978: 169). This is also a reflection of the inability to effectively expand both the domestic capital and intermediate goods industries in line with the increasing need of the industrialization process, although some efforts have been made towards this end through the heavy industry programme.

Malaysia's experience is similarly reflected in other developing countries of the region. Asian countries as a group continued to experience an increase in capital goods inflows during the 1980s, albeit at a slower rate in comparison with the 1970s. A study of 53 countries by the United Nations Conference on Trade and Development (UNCTAD, 1988) shows that for major exporters of manufactures, the growth rate per annum was positive at 4.1 per cent during the 1981–6 period, but one-fifth of the average for the 1970s, that is, at 21.7 per cent during the 1970–81 period (Table 2.10). Although the growth rate of capital goods inflows had generally fallen during the 1980s, it remained positive in the majority of countries in developing Asia and among those which were major exporters of manufactured goods. This was also reflected in the fact that the middle-income countries experienced a sharp drop in their imports of capital goods during the 1980s, while the higher-income developing countries recorded a moderate decline.

The relative decline in capital goods imports is the consequence of a number of factors which hindered the ability of developing countries to maintain their level of spending during the 1980s. There was a sharp deterioration in the export markets for developing countries, characterized principally by a fall in commodity prices in the early 1980s leading to lower output growth and thus constraining their spending on capital goods imports. At the same time, a considerable increase in interest rates world-wide added to the already heavy burden of debt servicing, while new borrowing which was required to meet debt service payments without lowering expenditures on capital goods and other development needs began to contract in the early 1980s.

# 3
# The Industrial Master Plan, 1986–1995

## The Industrial Master Plan and Its Objectives

THE IMP, launched in early 1986, was aimed at refocusing industrial planning from a largely market-oriented approach to a distinctly planned or target-oriented approach within a free enterprise economy. It can be argued that the former approach

> had served the country well in the early stages of Malaysia's drive to industrialise, but as the manufacturing sector was called upon to play an increasingly important role in spearheading the expansion of the Malaysian economy and as the process of industrialisation become more difficult, the Government decided that an Industrial Master Plan should be prepared for the country based on a plan-oriented approach to industrial planning as practised in Japan and Korea. Like in these countries, however, the market will continue to play a vital role in the economy and will ultimately ensure rational decisions involving resource allocation (Sadasivan, 1986).

However, more significantly, the launching of the IMP coincided with the state's concern about the lacklustre investment flows during the early 1980s, making it imperative for the reformulation of an industrial strategy to ensure greater investments in the manufacturing sector. According to the Fifth Malaysia Plan, overall foreign paid-up capital declined during the 1981–5 period due 'principally to adverse world economic conditions, uncertainties in the export market, and general tightening of international financial resources' (Malaysia, 1986: 340). At the same time, the total proposed capital investments approved by the Malaysian Industrial Development Authority (MIDA) during the period declined during 1983 and 1984 compared to 1982, before picking up again in 1985 and 1986 (Table 3.1).

On closer examination, one observes that the rate of increase for approved investments during the 1983–5 period was much slower for the foreign proportion compared to their local counterparts. In fact, the rate of increase of foreign investment was 14.1 per cent for 1984 and 33.6 per cent for 1985 while the respective rates for local investments

TABLE 3.1

Proposed Capital Investment and Employment Potential in Approved Projects, 1981–1989

| | *1981* | *1982* | *1983* | *1984* | *1985* | *1986* | *1987* | *1988* | *1989* | *Cumulative 1986–9* |
|---|---|---|---|---|---|---|---|---|---|---|
| Proposed capital investment ($ million) | 4,448.4 | 5,434.8 | 2,359.1 | 3,801.1 | 5,686.9 | 5,163.2 | 3,933.9 | 9,093.9 | 12,108.1 | 30,299.1 |
| Local | 3,139.1 | 3,808.2 | 1,729.0 | 3,083.1 | 4,727.6 | 3,475.3 | 1,873.9 | 4,215.9 | 3,540.1 | 13,105.2 |
| Foreign | 1,309.3 | 1,626.2 | 629.1 | 718.0 | 959.3 | 1,687.9 | 2,060.0 | 4,878.0 | 8,568.0 | 17,193.9 |
| Share of foreign investment (%) | 29.4 | 29.9 | 26.7 | 18.9 | 16.9 | 32.7 | 52.4 | 53.6 | 70.8 | 56.7 |
| Potential employment | 56,636 | 37,663 | 43,537 | 56,831 | 53,597 | 40,230 | 59,779 | 136,647 | 165,888 | 402,544 |

*Source*: MIDA (adapted from Low Peng Lum, 1990).

were 78.3 per cent and 53.3 per cent. An interesting point is that while foreign investments increased by 75.9 per cent in 1986, the local proportion decreased by 26.5 per cent in the same year thus increasing the share of foreign investments in approved projects to a stronger position of 32.7 per cent. A similar trend was to be observed in the following years, which, as shall be examined in the following chapters, ironically poses a predicament for domestic manufacturing industries with respect to their technological dependence on foreign capital.

While the principal rationale of the IMP should be understood, it is also important that its objectives are noted to assess its general impact on the overall development of the manufacturing sector, and more specifically on domestic technological development. The objectives of industrial development set by the Plan are:

1. To accelerate the growth of the manufacturing sector to ensure a continued rapid expansion of the economy and to provide a basis for meeting the social objectives consistent with the New Economic Policy.
2. To promote opportunities for the maximum and efficient utilization of the nation's abundantly endowed natural resources.
3. To build up the foundation for leap-frogging towards advanced industrial country status in the information age by increasing indigenous technological capability and competitiveness.

To achieve all the three main objectives, the IMP views the macroeconomic prospects of the Malaysian economy up to the year 1995 as optimistic. The growth rate of the GDP during the Plan period is expected to be 6.4 per cent per annum, and to support this GDP growth the total investment for the economy should increase at the rate of 5.7 per cent annually in real terms. The IMP also projects that the growth rate of the manufacturing sector will be 8.8 per cent per annum in real terms so that by 1995 the share of manufacturing in GDP will rise to 23.9 per cent from 19.1 per cent in 1985 and 21.7 per cent in 1990. During the Plan period, the manufacturing sector is also expected to create substantial employment opportunities, amounting to an additional 705,400 new jobs. If the employment potential indicated by new investment projects were to be realized, this target seems to be achievable during the remaining years of the Plan period. As indicated in Table 3.1, the employment potential during the 1986–9 period is 402,544 (that is, 57 per cent of the IMP employment target), and this does not include those to be created by manufacturing projects not subject to government approval under the Industrial Coordination Act, 1975.

With respect to the critical development of technology, the third objective seems to be very crucial although the second objective tacitly implies the need for technological and skill enhancement particularly in resource-based industries. In terms of the third objective, the IMP appropriately emphasized the following:

> To attain the status of an industrial country in the future, Malaysia should devote substantial efforts to achieve a high degree of efficiency in manufacturing

activities. The objective of manufacturing development should thus be oriented towards increasing efficiency. In particular, efficiency requires the following: competitive markets; entrepreneurial spirits; an achievement orientation; and the accumulation of skills and technologies (UNDP/UNIDO, 1985: 52).

The launching of the IMP was subsequently complemented by the introduction of the Promotion of Investments Act in 1986 and the amendment of the Income Tax Act of 1967 to provide liberal investment incentives to potential investors. Although it has been the policy of the state to encourage projects on a joint-venture basis with local capital, especially those which are largely dependent on the domestic market, the introduction of the 1986 Act allowed more flexibility on the part of the authorities to approve higher foreign equity participation in any manufacturing project. Among the guidelines introduced are the following:

1. Foreign investors are allowed to hold up to 100 per cent equity in a firm if the latter exports 80 per cent or more of its production.
2. For firms exporting between 51 and 79 per cent of their production, foreign equity ownership up to 51 per cent will be allowed. However, foreign equity ownership up to 79 per cent may be allowed, depending on factors such as the level of technology, spin-off effects, size of the investment, location, value-added, and the utilization of locally produced raw materials and components.
3. For firms exporting between 20 and 50 per cent of their production, foreign equity ownership of between 30 per cent and 51 per cent will be allowed depending upon similar factors as mentioned above. However, for firms exporting less than 20 per cent of their production, foreign equity ownership is allowed up to a maximum of 30 per cent.
4. For firms producing products which are highly technology-intensive and regarded as 'priority products' for the domestic market, foreign equity ownership up to 51 per cent will also be allowed.

Apart from the above, more relaxations were made in October 1986 allowing foreign investors (whose applications were received during the period 1 October 1986 to 31 December 1990) to hold up to 100 per cent equity if the following conditions are met:

1. If the firm exports 50 per cent or more of its production, the firm employs 350 full-time Malaysian workers, and such employment at all levels reflects the racial composition of the country.
2. That the firm's products do not compete with products presently being manufactured locally for the domestic market.

The effects of these changes have culminated in increased flows of direct foreign investment particularly since 1987. Although the proportion of foreign proposed capital investment to the total declined steadily during the 1980–5 period as indicated in Table 3.1, it has increased quite rapidly since then, reaching a peak of 70.8 per cent in 1989. The percentage of wholly foreign-owned projects to the total also increased from 4.8 per cent in 1985 to 24.9 per cent in 1987 and peaked at 35.2 per cent in 1989. The percentage of approved projects with a foreign majority also increased from 7.7 per cent in 1985 to 15.4 per cent

TABLE 3.2
Projects Granted Approval by Ownership, 1981–1989

| *Ownership* | *Number of Projects* | | | | | | | | |
|---|---|---|---|---|---|---|---|---|---|
| | *1981* | *1982* | *1983* | *1984* | *1985* | *1986* | *1987* | *1988* | *1989* |
| Wholly Malaysian-owned | 293 | 260 | 244 | 380 | 321 | 177 | 105 | 262 | 184 |
| | (49.2) | (55.6) | (49.8) | (50.7) | (51.4) | (39.6) | (31.6) | (35.8) | (23.5) |
| Wholly foreign-owned | 27 | 16 | 24 | 24 | 30 | 30 | 83 | 169 | 276 |
| | (4.5) | (3.4) | (4.9) | (3.2) | (4.8) | (6.7) | (24.9) | (23.1) | (35.2) |
| Joint ventures | | | | | | | | | |
| Malaysian majority | 184 | 140 | 172 | 276 | 214 | 158 | 86 | 141 | 148 |
| | (30.9) | (29.9) | (35.1) | (36.9) | (34.2) | (35.3) | (25.8) | (19.3) | (18.9) |
| Foreign majority | 84 | 47 | 45 | 61 | 48 | 69 | 49 | 132 | 156 |
| | (14.1) | (10.0) | (9.2) | (8.1) | (7.7) | (15.4) | (14.7) | (18.0) | (19.9) |
| Equal ownership | 8 | 5 | 5 | 8 | 12 | 13 | 10 | 28 | 20 |
| | (1.3) | (1.1) | (1.0) | (1.1) | (1.9) | (2.9) | (3.0) | (3.8) | (2.5) |
| Total | 596 | 468 | 490 | 749 | 625 | 447 | 333 | 732 | 784 |
| | (100.0) | (100.0) | (100.0) | (100.0) | (100.0) | (100.0) | (100.0) | (100.0) | (100.0) |

*Source*: MIDA (quoted in Low Peng Lum, 1990).
*Note*: Figures in parentheses indicate percentage of total.

in 1986 and to 19.9 per cent in 1989 (Table 3.2). It must, however, be noted that the total amount of capital investment contributed by local investors is not fully reflected in Table 3.1, particularly so after 1987 when the Industrial Coordination Act of 1975 was revised and no government approval was needed for investment projects with shareholders' funds of less than $2.5 million or with less than 75 full-time employees.

At the same time, approvals granted for the establishment of highly capital-intensive projects could have distorted the capital investment figures during these years. For instance, in 1988 an unwrought aluminium project approved for location in Sarawak involved a proposed foreign capital investment of $525 million while a $500 million ethylene and propylene project with 100 per cent foreign ownership was approved. In 1989, a methyl tertiary butylether/propylene project with a proposed foreign capital investment of $216 million and a $240 million cathode ray tube project with 100 per cent foreign ownership were approved (Low Peng Lum, 1990).

## 'Leap-frogging' into an Industrialized Economy

Obviously concerned with the prospect that the technological gap between Malaysia and the newly industrialized countries, not to mention the advanced industrial countries, may widen into a chasm, the IMP suggests that the best hope for catching up is to 'leap-frog'. However, the term 'leap-frogging' has been used to depict a number of different circumstances which has often led to numerous ambiguities about its meaning. In its most common use, it refers to efforts initiated by an economy or the manufacturing sector to surpass the existing state of the art in the development of new technologies which are still in their pre-standardization or pre-commercialization stage. Despite the high risks and the substantial investments in resources needed in this kind of endeavour, the propensity to catch up would be greater if such efforts were made at an early stage, that is, before they have been standardized and commercialized. This would essentially require considerable amounts of complementary inputs that are critical for ensuring success, such as an abundant supply of skilled manpower and considerable R & D activities.

Another scenario for industrial 'leap-frogging' could be associated with efforts to enter a new technology system when it is at an even more basic exploratory stage, that is, when the area is scientifically and technically well endowed but the commercial potential is largely unknown. The risks are even higher in this particular case compared to the previous case, where the commercial potential is more apparent and the time frame for commercial developments is short. The main constraint would be the mobilization of all the complementary inputs in order to exploit their commercial potential, for example, advanced material handling and fabrication technology, process engineering know-how, and product design and development capabilities.

The emergence of a number of South Korean firms in the fabrication of large computer memories (d-RAMs), for instance, has been referred to as 'leap-frogging'. Although this industry has been characterized by high barriers to entry, basically because of the large capital outlays, the *chaebols* were perhaps pre-eminently positioned to reap the economies of scale. The support of the state and effective planning at the firm level were crucial elements in their commercial success, but they are continually faced with escalating R & D and manpower requirements if they are to stay competitive *vis-à-vis* their competitors in Japan or the United States. Yu (1989: 337–57), for example, emphasized three factors contributing to the commendable performance of the South Korean electronics industry:

> First, the government and the industry have, either jointly or separately, monitored the changes in the industry and have updated the development plans to maintain its applicability. This process had promptly singled out those policies and incentives that need to be adjusted and rectified. The second factor is the human factor. There were: (1) a sufficient supply of technicians and engineers who acted as technology absorbing and adapting agents, (2) a large number of entrepreneurs who were willing to take risks, and (3) a competitive domestic market environment where the entrepreneurs worked hard to survive before venturing into the export market. The third and most important factor was the business enterprise itself. Business firms like Samsung and Goldstar resolved all their problems on their own, which utilised the entrepreneurial spirit both innovatively and aggressively.

In contrast to the above cases, the main concern of the IMP has been the ability to 'leap-frog' towards advanced industrial country status in which the eminence of 'information' is the most critical determinant of a country's technological development, and this arises from the extremely rapid expansion of information technology during the 1970s and 1980s as a result of successive refinements and innovative efforts in computer and micro-electronics technologies. The dramatic changes that have taken place in electronics and telecommunications, and their effects, illustrate many of the traits of new technologies. Apart from their impact on the processes and products within the electronics and telecommunications sectors, these developments created new possibilities in a wide range of sectors, including process control in the more critical industries such as steel and petrochemicals, automation in assembly-type industries such as motor vehicles and other consumer durables, and automated processing and communications in the banking and insurance industries.

The availability of computer-based and communications-linked information processing systems, complemented by their effectiveness in cost reduction and productivity enhancement across a broad spectrum of industries and applications, has rendered their adoption virtually a necessity if an industry wishes to compete successfully. Thus, the failure, or even the lack of haste, to adopt such technologies may create a further widening of the already formidable technological gap separating developing countries (such as Malaysia) and the more industrialized countries. The enhancement of the country's information resources must therefore be accorded the highest priority. This is especially significant given the export-orientation of the country's industrialization effort since the mid-1980s. It is in this light and that of objective (3) of the IMP that the state initiated in 1988 the formulation of an Action Plan for Industrial Technology Development.

One of the many critical constraints identified in the IMP is the weak inter-industry linkages within manufacturing, arising principally from the narrow industrial base which is sustained by a few export-oriented and agro-resource-based industries characterized by relatively low levels of technology. (The significant ones are electronics, textiles, food products, wood-based products, and rubber products.) While a major proportion of the electronics and textile production originates from the FTZs and is almost exclusively produced by foreign multinational corporations providing minimal linkages to the rest of the economy, the resource-based industries are generally concentrated in primary processing and have yet to exploit their potential in terms of downstream activities.

At the same time, intersectoral linkages are equally weak. This is particularly so between the manufacturing sector and the primary sector (that is, agriculture and mining) where agriculture (including forestry and fisheries) still plays a significant role in terms of output and employment. Despite being relatively successful in diversifying its economy, Malaysia has been unable to substantively increase the economic linkages between industry and the primary sector, the latter having been the mainstay of the economy for many decades. As a result, many of the industries that are established have remained peripheral to the interests of the primary sector, mainly agriculture.

The agricultural sector will undoubtedly remain significant during the 1990s despite the structural changes envisaged by the IMP. During this decade, such changes will not be so dramatic as to change the role of the agricultural sector critically, particularly with respect to employment for the rural population. While new industrial strategies will be implemented and a new direction in industrial development sought, the development of the agricultural sector will nevertheless be given emphasis in macroeconomic planning as reflected in the National Agricultural Policy. This Policy, introduced in 1984, was aimed at the modernization and revitalization of the sector so as to maximize income through the efficient utilization of resources (Malaysia, 1986: 296). It will be even more advantageous if agricultural development is effectively linked to the expansion of the industrial sector, whose growth is anticipated to provide the catalyst for overall development during the 1990s. The Mid-Term Review of the Fifth Malaysia Plan (1986–90) appropriately emphasized that:

Efforts at diversification will remain including development of processing and value added activities through linkages with the manufacturing sector to ensure a stable growth of the agriculture sector and provide for a balanced source of income generation for farmers as a strategy to overcome rural poverty as promulgated in the New Economic Policy (Malaysia, 1989a: 161).

An important component with regard to the creation of these linkages, at least in terms of policy objective, is rural industrialization. Although much has been said about this strategy in the previous Five-year Plans, and reiterated in the Sixth Malaysia Plan (Malaysia, 1991b: 148), there is still a gap between policy aspiration and policy implementation. At the present stage of the country's industrialization process in the early 1990s, firms prefer to locate their plants in the more developed urban centres for reasons of agglomeration and infrastructure; the institutional mechanism to encourage the growth of industries in the rural areas is still relatively weak.

## The Development of Resource-based Industries

One of the principal strategies of the IMP is the intensive development of resource-based industries (RBIs), particularly those sub-sectors that have the potential for export promotion and for enhancing technical skills. Seven industries have been identified as RBIs: (i) rubber products,

TABLE 3.3
Average Annual Production Growth Rate of Manufacturing Industries, 1971–1983 (percentage)

| *Industry* | *1971–80* | *1976–80* | *1981–3* |
|---|---|---|---|
| Food processing | 4.5 | 6.9 | −0.3 |
| Oils and fats | 20.5 | 15.4 | 9.1 |
| Other foods | 4.2 | 6.3 | 4.8 |
| Beverages and tobacco | 8.2 | 9.5 | −3.3 |
| Textiles and clothing | 12.6 | 11.9 | −2.6 |
| Sawmills and furniture | 8.6 | 8.6 | 0.4 |
| Paper and printing | 11.9 | 16.4 | 6.2 |
| Chemical products | 6.7 | 10.8 | −11.6 |
| Petroleum products | 6.9 | 10.4 | 18.6 |
| Rubber processing | −3.3 | −0.5 | −1.9 |
| Rubber products | 5.0 | 4.4 | −3.3 |
| Cement | 8.6 | 10.1 | 10.7 |
| Other non-metallic products | 9.9 | 15.4 | −4.4 |
| Basic metal products | 9.9 | 9.5 | 3.2 |
| Fabricated metal products | 9.4 | 11.7 | 3.8 |
| Electrical machinery | 11.2 | 11.0 | 10.7 |
| Transport equipment | 12.1 | 15.6 | 4.9 |
| Total manufacturing | 11.4 | 11.3 | 4.4 |

*Source*: Malaysia (1984).

(ii) palm oil products, (iii) food processing, (iv) wood-based products, (v) chemical and petroleum products, (vi) non-ferrous metal products, and (vii) non-metallic mineral products.

Although Malaysia has a comparative advantage in most of these industries, their exports are principally characterized by primary processing and low value-added products so that their contribution to export promotion is still minimal. As shown in Table 3.3, on average the RBIs have been growing at a slower rate than the overall manufacturing industry during the 1971–83 period. The major exceptions were oils and fats and paper and printing whose average annual growth rates were 20.5 per cent and 11.9 per cent respectively during the 1970s (well above the manufacturing sector's average of 11.4 per cent), while during the 1981–3 period, with the overall decline in manufacturing growth, these two industries sustained a relatively high growth rate.

Other RBIs, particularly petroleum products and the cement industry, experienced relatively high annual growth rates (at 18.6 per cent and 10.7 per cent respectively during the 1981–3 period) although industries like food processing, chemical products, rubber products and processing, and other non-metallic products indicated negative growth during the same period. During the 1986–8 period, the rubber processing and rubber products industries were the only RBIs achieving commendable growth in comparison with overall manufacturing growth (Table 3.4). During the 1989–90 period, these two industries are expected to sustain their rapid growth, while other resource-based industries such as food processing, oil and fats, and non-metallic mineral products are expected to experience reasonable rates of growth.

Despite all these changes, there seems to be a general decline in the overall contribution of the RBIs towards total manufacturing output. Table 3.5 indicates that, with the exception of food processing (including beverages and tobacco), paper and paper products, chemical products, and non-metallic mineral products which were able either to sustain or to increase marginally their share of value-added in total manufacturing, the contribution of the RBIs had generally declined during the 1973–86 period. The electrical machinery industry (particularly the electronic and electrical products sub-sectors) had, on the other hand, shown a tremendous increase in its value-added share during this period. This change in the industry-mix may be associated with the relative lack of emphasis on the RBIs, although more significantly, the growth of non-RBIs tended to be faster as the phase of industrialization shifted towards export-oriented and labour-intensive lines of production, particularly since the mid-1970s.

If priority is then given to such industries (selecting the most suitable product lines), the industrialization process in the 1990s could be productively linked to the development of the agricultural sector as the main supplier of inputs, thus promoting 'backward linkages' as far as manufacturing is concerned. This will specifically relate to industries such as rubber products, palm oil products, food processing, and wood-based products.

TABLE 3.4
Manufacturing Production Index, 1986–1990 (1985 = 100)

| Industry | Weights | Production Index | | | | | | Annual Growth Rate (%) | | | | | Average Annual Growth Rate (%) | | |
|---|---|---|---|---|---|---|---|---|---|---|---|---|---|---|---|
| | | *1985* | *1986* | *1987* | *1988* | *1989* | *1990* | *1986* | *1987* | *1988* | *1989* | *1990* | *1986–8* | *1989–90* | *1986–90* |
| Food manufacturing | 4.3 | 100.0 | 102.2 | 114.0 | 121.9 | 142.6 | 155.5 | 2.2 | 11.6 | 6.9 | 17.0 | 9.0 | 6.8 | 12.9 | 9.2 |
| Oils and fats | 4.8 | 100.0 | 114.6 | 117.8 | 122.3 | 152.8 | 166.6 | 14.6 | 2.8 | 3.8 | 25.0 | 9.0 | 6.9 | 16.7 | 10.7 |
| Beverages | 1.5 | 100.0 | 100.5 | 104.6 | 117.3 | 129.0 | 138.1 | 0.5 | 4.1 | 12.2 | 10.0 | 7.0 | 5.5 | 8.5 | 6.7 |
| Tobacco manufacturing | 2.7 | 100.0 | 79.2 | 79.3 | 91.6 | 100.8 | 109.8 | – 20.8 | 0.1 | 15.5 | 10.0 | 9.0 | –2.9 | 9.5 | 1.9 |
| Textiles and clothing | 3.0 | 100.0 | 113.5 | 131.6 | 133.3 | 156.0 | 179.4 | 13.5 | 15.9 | 1.3 | 17.0 | 15.0 | 10.1 | 16.0 | 12.4 |
| Wood and cork products (except furniture) | 3.4 | 100.0 | 96.6 | 116.6 | 132.4 | 143.0 | 153.0 | –3.4 | 20.7 | 13.6 | 8.0 | 7.0 | 9.8 | 7.5 | 8.9 |
| Industrial chemicals and other chemical products | 10.0 | 100.0 | 115.7 | 129.9 | 138.9 | 151.4 | 160.5 | 15.7 | 12.3 | 6.9 | 9.0 | 6.0 | 11.6 | 7.5 | 9.9 |
| Petroleum refineries | 1.8 | 100.0 | 121.8 | 127.6 | 135.5 | 143.6 | 150.8 | 21.8 | 4.8 | 6.2 | 6.0 | 5.0 | 10.7 | 5.5 | 8.6 |

| | | | | | | | | | | | | | | | |
|---|---|---|---|---|---|---|---|---|---|---|---|---|---|---|---|
| Rubber remilling and latex processing | 1.1 | 100.0 | 107.8 | 128.8 | 165.3 | 239.7 | 275.6 | 7.8 | 19.5 | 28.3 | 45.0 | 15.0 | 18.2 | 29.1 | 22.5 |
| Rubber products | 2.2 | 100.0 | 118.5 | 145.4 | 232.6 | 337.3 | 455.3 | 18.5 | 22.7 | 60.0 | 45.0 | 35.0 | 32.5 | 39.9 | 35.4 |
| Non-metallic mineral products | 3.9 | 100.0 | 84.1 | 78.8 | 92.8 | 116.0 | 148.5 | −15.9 | −6.3 | 17.8 | 25.0 | 28.0 | −2.5 | 26.5 | 8.2 |
| Iron and steel and non-ferrous metals | 2.4 | 100.0 | 84.5 | 101.1 | 126.5 | 145.5 | 171.7 | −15.6 | 19.9 | 24.9 | 15.0 | 18.0 | 8.2 | 16.5 | 11.4 |
| Fabricated metal products (except machinery and equipment) | 1.9 | 100.0 | 94.9 | 111.4 | 118.8 | 130.7 | 139.8 | − 5.1 | 17.4 | 6.7 | 10.0 | 7.0 | 5.9 | 8.5 | 6.9 |
| Electrical machinery, appliances, and supplies | 9.6 | 100.0 | 141.6 | 172.9 | 209.6 | 226.4 | 237.7 | 41.6 | 22.1 | 21.3 | 8.0 | 5.0 | 28.0 | 6.5 | 18.9 |
| Transport equipment | 2.7 | 100.0 | 73.5 | 75.6 | 132.8 | 139.4 | 149.2 | −26.5 | 2.8 | 75.7 | 5.0 | 7.0 | 9.9 | 6.0 | 8.3 |
| Total manufacturing | 55.3 | 100.0 | 109.4 | 123.7 | 145.8 | 165.5 | 183.7 | 9.4 | 13.0 | 17.9 | 13.5 | 11.0 | 13.4 | 12.2 | 12.9 |

*Source*: Malaysia (1989a).

TABLE 3.5
Value-added Share of the Manufacturing Sector, 1973–1986 (percentage)

| *Industry* | *1973* | *1978* | *1983* | *1986* |
|---|---|---|---|---|
| Food, beverages, and tobacco | 17.2 | 15.8 | 16.2 | 17.0 |
| Palm oil and palm kernel oil | 4.7 | 10.2 | 9.2 | 5.6 |
| Textiles and wearing apparel | 5.2 | 7.9 | 5.5 | 7.0 |
| Leather and footwear | 0.1 | 0.1 | 0.1 | 0.2 |
| Wood and wood products, furniture and fixtures | 15.6 | 10.4 | 5.8 | 5.7 |
| Paper and paper products, printing and publishing | 5.3 | 4.8 | 5.1 | 5.1 |
| Chemical products | 7.1 | 5.7 | 5.9 | 6.6 |
| Petroleum refineries | 3.8 | 3.3 | 2.4 | 3.5 |
| Rubber products | 8.0 | 9.9 | 6.5 | 7.5 |
| Plastic products | 2.6 | 1.8 | 2.0 | 2.2 |
| Non-metallic mineral products | 4.9 | 5.1 | 5.0 | 6.0 |
| Basic metal products | 5.9 | 3.1 | 5.2 | 3.8 |
| Fabricated metal products | 4.9 | 3.8 | 4.1 | 3.0 |
| Machinery | 3.3 | 2.9 | 3.6 | 2.3 |
| Electrical machinery | 7.3 | 10.8 | 17.4 | 17.6 |
| Transport equipment | 3.3 | 3.0 | 4.2 | 3.1 |
| Others | 0.8 | 1.3 | 1.8 | 3.8 |
| Total manufacturing (%) | 100.0 | 100.0 | 100.0 | 100.0 |

*Sources*: Computed from *Industrial Surveys 1973*, *1978*, *1983*, and *1986*, Department of Statistics, Malaysia.
*Note*: Value-added in 1978 constant prices.

Within the framework of achieving the targeted growth rates during the 1986–95 period and the need to have an export-oriented economy, the IMP also explicitly projects that the leading industries will be electronics, machinery, and textiles; each industry is projected to achieve growth rates exceeding 9 per cent per annum. (Additional industries that show some prospects for future expansion include oleochemicals, tyres, furniture, cement, and petrochemical and petrochemical products.) In terms of employment generation, the three leading industries and the transport equipment industry will amongst them provide about 53 per cent of the total manufacturing employment (amounting to 705,400 new jobs) during this period, bringing the number involved in manufacturing to 1.4 million people. If the wood product and basic metal product industries are included, the percentage of new employment created by these six industries will amount to 72 per cent of total employment in manufacturing.

In terms of export-orientation, the IMP, according to the Ministry of International Trade and Industry, has successfully created a favourable competitive position for the export of manufactures in overseas markets.

This competitive edge, coupled with a strong external demand, has caused a sharp upsurge in manufactured exports since the implementation of the Plan in 1986. During 1989, for example, export of manufactures, comprising 54.7 per cent of the country's total exports, increased by 34.2 per cent; thus, for a third consecutive year growth of manufactured exports exceeded 30 per cent (Mohd Yusof Ismail, 1989). Looking at the important industry sectors earmarked for export growth (Table 3.6), the targets set by the IMP have generally been achieved by almost all sectors, particularly in 1988. With the exception of the food processing, chemical, non-metallic mineral products, and processed palm oil sectors, other sectors recorded excellent export growth, particularly the electrical and electronics, textile and garments, and iron and steel sectors which exceeded their respective IMP targets. Again, it must be noted that a substantial influence in this export expansion has been the multinational corporations located in the FTZs, rather than any increase in domestic manufacturing enterprises.

For example, actual exports from the electrical and electronics sector consistently exceeded the Plan's export targets: by 44.6 per cent in 1986, 62.1 per cent in 1987, and 101.6 per cent in 1988 for the electronics sub-sector; and by 77.0 per cent, 127.1 per cent, and 215.1 per cent respectively for the electrical sub-sector. The initiatives taken to promote export-oriented industrialization have therefore led to the rapid growth of the manufacturing sector, and judging by the composition of new investments approved during the 1986–8 period, export-led growth is expected to continue as the major thrust of the country's industrialization efforts. Table 3.7, for instance, indicates that export-oriented projects accounted for 78.9 per cent of the total capital investments amounting to $9,093.9 million approved in 1988 compared to 38.4 per cent in 1985 and 40.2 per cent in 1980.

As indicated earlier, the second principal objective of the IMP is to promote opportunities for the maximum and efficient utilization of the country's abundant resources. Amongst the industrial strategies adopted to achieve this objective are the diversification of export markets, the modernization of the small- and medium-scale sector and the rationalization of inefficient and declining firms, the development of indigenous technology and technological absorptive capability, the expansion and strengthening of manpower, the integration of information services, and the extensive application of informatics. Given the above-mentioned principal objective and the diversity of strategies proposed (the strategies may even conflict with one another), the obvious question to ask is, how should 'industry and product' that are to be prioritized be ranked within manufacturing? Such a priority ranking is indeed imperative because of the competition for scarce resources, both in terms of capital and manpower. It is also important as it affects the degree of emphasis given to the numerous strategies to be adopted during the IMP period as well as policies related to industrial technology development.

If the main objective of the IMP is to maximize the efficient utilization

TABLE 3.6
Export Performance against IMP Targets: 1986, 1987, and 1988 ($ million)

| Industry | 1986 | | | 1987 | | | 1988 | | |
|---|---|---|---|---|---|---|---|---|---|
| | *IMP Target* | *Actual* | *% Change* | *IMP Target* | *Actual* | *% Change* | *IMP Target* | *Actual* | *% Change* |
| Resource-based Industries | | | | | | | | | |
| Rubber | 329.5 | 397.9 | 20.8 | 424.4 | 581.1 | 35.6 | 478.4 | 1,039.6 | 117.3 |
| Wood | 1,779.9 | 1,655.0 | –25.9 | 2,018.8 | 2,633.6 | 30.5 | 2,137.7 | 3,200.7 | 49.7 |
| Food processing | 915.1 | 895.7 | –2.1 | 1,024.2 | 1,073.1 | 4.8 | 1,153.6 | 1,257.2 | 9.0 |
| Chemicals | 379.1 | 555.1 | 46.4 | 429.2 | 592.4 | 38.0 | 476.4 | 896.2 | 88.1 |
| Non-ferrous metal | | | | | | | | | |
| Tin | 15,086.0 | 18,300.0 | 21.3 | 16,677.0 | 34,200.0 | 105.3 | 18,200.0 | 46,500.0 | 155.5 |
| Aluminium | 15,646.0 | 14,700.0 | 169.7 | 17,149.0 | 70,100.0 | 308.7 | 17,800.0 | 86,100.0 | 383.7 |
| Copper | 2,974.0 | 13,600.0 | 357.3 | 3,407.0 | 45,800.0 | 1,244.3 | 3,600.0 | 111,600.0 | 3,000.0 |
| Non-resource-based | 210.7 | 152.7 | –27.5 | 239.6 | 224.1 | –6.5 | 250.0 | 370.0 | 48.0 |
| Non-resource-based Industries | | | | | | | | | |
| Electrical and Electronics | | | | | | | | | |
| Electrical | 283.0 | 501.0 | 77.0 | 330.0 | 750.0 | 127.1 | 357.0 | 1,125.0 | 215.1 |
| Electronics | 4,828.0 | 6,979.0 | 44.6 | 5,830.0 | 9,450.0 | 62.1 | 6,507.0 | 13,118.0 | 101.6 |
| Transportation | n.a. | n.a. | n.a. | n.a. | n.a. | n.a. | – | 67.3 | – |
| Machinery and engineering | 443.0 | 826.6 | 87.0 | 481.0 | 1,218.3 | 153.0 | 538.0 | 1,668.4 | 210.0 |
| Iron and steel | 38.2 | 250.0 | 553.0 | 43.6 | 391.0 | 796.0 | 42.3 | 498.0 | 1,077.0 |
| Textiles and garments | 934.1 | 1,655.2 | 77.2 | 1,053.0 | 2,279.0 | 116.5 | 1,200.8 | 3,044.6 | 153.5 |

*Source*: Ministry of Trade and Industry (quoted in Mohd Yusof Ismail, 1989).
n.a. = Not available.

TABLE 3.7
Approved Projects According to Numbers and Capital Investment, 1980, 1985, and 1988

| | *Export-oriented Projects*[a] | | *Total Manufacturing Projects* | |
|---|---|---|---|---|
| *Year* | *Number* | *Capital Investment ($ million)* | *Number* | *Capital Investment ($ million)* |
| 1980 | 156 | 844.9 | 459 | 2,102.8 |
| | (33.9) | (40.2) | (100) | (100) |
| 1985 | 147 | 2,184.9 | 625 | 5,686.9 |
| | (23.5) | (38.4) | (100) | (100) |
| 1988 | 618 | 7,183.2 | 732 | 9,093.9 |
| | (84.4) | (78.9) | (100) | (100) |

*Source*: MIDA (quoted in Leong, 1989).
*Note*: Figures in parentheses are percentages of the total.
[a]Projects exporting 50 per cent or more of their output.

of the country's natural resources, it is certainly consistent with the aim to develop more intersectoral linkages as examined in this chapter. Looking at the existing structure of the manufacturing sector, four major industry groups within the RBIs can be given emphasis for industrial deepening. These industry groups include food processing (particularly oils and fats), wood-based products, rubber products, and petroleum products. The emphasis on these industries would also be consistent with the objective of expanding and diversifying domestic manufacturing activities so as to increase each industry's value-added component. Since these industries have important inter-industry linkages within the economy through the agricultural sector and the country's natural resources, it is imperative that their potential be fully developed to exploit the country's comparative advantage. This could be particularly significant for industries producing oils and fats, wood-based products, and rubber products in the short and medium term, and for the petroleum products industry in the long term.

Having identified the main RBIs to be promoted, it is important to highlight the major strategies proposed and examine their implications for the utilization of domestic resource endowments. Amongst these strategies, two special areas can be identified that would have direct relevance on the promotion of these industries and to the creation of greater economic linkages with the agricultural sector. These are related to the development of indigenous technology and technological absorptive capacity as well as the development and modernization of the small- and medium-scale industries. The former requires greater emphasis on measures that would lead to technological enhancement within domestic industries.

This does not imply that the other strategies proposed by the IMP are unimportant. However, there is the need to put these strategies in their proper perspective so as not to lose sight of the fact that Malaysia, being a developing country in which the industrial base is relatively small, requires a more efficient means of allocating its capital and manpower resources. This also implies that the industrial strategies to be implemented during the IMP period have to be prioritized to achieve the desired objectives. It is in this sense that technology development, including domestically initiated R & D activities, and the modernization of small- and medium-scale industries need immediate focus during the implementation of the IMP. On the other hand, other IMP strategies should not be neglected but given emphasis within a stipulated time frame, taking into consideration the institutional constraints within the economy. For example, as shall be seen in the following chapters, the emphasis on R & D must also be related to policies regarding human resource development although the impact of the latter is only felt in the long term rather than in the immediate and medium term.

Owing to a number of constraints (including the needs of the agricultural sector), Malaysia currently is unable to make industrial R & D its strategic priority. During the 1980s the Standards and Industrial Research Institute of Malaysia (SIRIM) was the only public sector research institute which was actively involved in this area besides the Malaysian Institute of Microelectronic Systems (MIMOS) which was established in 1985, while other public sector research institutions such as the Rubber Research Institute of Malaysia (RRIM), the Palm Oil Research Institute of Malaysia (PORIM), and the Malaysian Agricultural Research and Development Institute (MARDI) are directly involved in agriculturally oriented research. The private sector's contribution in this respect is comparatively small, and thus its impact is minimal. While public sector R & D activities in agriculture continue to be dominant, the Sixth Malaysia Plan emphasizes that there is now a significant shift towards more downstream application-oriented research in the processing of agricultural output and by-products. In comparison to agriculture, however, R & D activities in industry have not been given as much emphasis mainly because of the narrow base of the sector itself and existing constraints in technology and manpower (Malaysia, 1991b: 189–90).

This predicament becomes more complicated in view of the general reluctance on the part of private industry to invest in R & D facilities because of the risks involved, but partly also because of the lack of highly skilled manpower. This is particularly obvious in the case of domestic firms which are relatively small and therefore lacking in the financial and manpower resources to initiate any kind of R & D activity. Foreign-owned or -managed enterprises (with the exception of a very few) are also reluctant to develop extensive R & D facilities locally for different reasons—they will prove wasteful in terms of their global interests. This is logical from their point of view considering that they have already established such R & D facilities within their parent companies. The lack of qualified R & D manpower and the absence of a tradition to innovate make it less attractive for them to site these facilities in developing countries such as Malaysia. If any R & D facility has been established by them, it is generally confined to testing and quality control activities rather than oriented towards product or process development.

## The Industrial Master Plan and the Critical Role of Technology

The IMP has identified twelve major industry sectors which will spearhead the country's industrialization programme in the 1990s. In this respect, it is important to note the major development objectives of the main industry sectors as far as they relate to technology development within domestic industries as a whole (Malaysia, 1987).

*Rubber Products*

1. To encourage aggressive export promotion and development of significant market niches of selected key rubber products, particularly tyres and latex-dipped goods, by improving the level of competitiveness through adopting cost-reduction measures and increasing productivity and product quality in order to meet the IMP export targets.
2. To encourage greater foreign investment, especially by multinational corporations (MNCs), in order to gain access to export markets and attain greater co-operation with them on R & D activities.
3. To place Malaysia in the forefront of R & D in rubber product manufacturing as well as in natural rubber production through more financial support to local and overseas institutions.

*Palm Oil Products*

1. To rationalize the palm oil refining industry and fractionation subsectors in order to increase their efficiency and competitiveness in the export markets.
2. To promote the oleochemicals as priority products and as leading export items.

*Food Processing Industry*

1. To extend the range of food products that could be import-substituted and to develop new product lines through effective R & D.
2. To seek ways and means of reducing production costs in the industry through innovative production processes.

*Wood-based Industry*

1. To rationalize inefficient and uneconomical sawmills and plywood mills through diversification into the manufacturing of downstream products or to merge them into larger, economic units.
2. To create a large production base through the establishment of furniture complexes and timber processing zones.
3. To develop the technological base of the industry through upgrading of design, product development, research into materials, and production technology.

*Chemical Industry*

1. To stimulate the demand for fertilizers by optimizing their usage in the agricultural sector.
2. To maximize the utilization of palm oil-based industries in order to make Malaysia a significant exporter of soaps and detergents.
3. To identify and exploit chemical deposits which can be tapped for production of inorganic chemicals.
4. To create a pool of local technical expertise to enable proper transfer and adaptation of acquired technology.

*Electronics and Electrical Industry*

For the electronics industry:

1. To foster the development of supplier and support industries.
2. To encourage the production of higher value-added products and R & D activities and to improve design capabilities within the sector.
3. To improve and upgrade the technology used in the existing semiconductor industry and to test activities in order to increase productivity.

For the electrical industry:

1. To enhance domestic technological capability through the encouragement of greater R & D activities, skill upgrading, and product development within the sector in order to lessen the dependence on foreign designs for the manufacture of electrical products.
2. To foster the development of efficient ancillary industries to support the growth of the industry.

*Transport Equipment Industry*

1. To restructure and rationalize the existing motor vehicle assemblers in order to achieve economies of scale and to develop the national car project as the focal point for the manufacture of component parts.
2. To formulate an effective local-content programme to ensure the industry's competitiveness.

*Machinery and Engineering Industry*

1. To upgrade the level of domestic technology in terms of both manufacturing and designing capabilities through the introduction of specific R & D and training programmes.
2. To expand the local supply capability by providing greater encouragement and incentives for the sub-sectors to modernize and diversify into a wider range of components and services of acceptable standards and quality.
3. To specifically promote the production of certain core products such as moulds, tools and dies, and castings and forgings which are the components used extensively in the manufacture of most machinery and equipment, and to create demand for the establishment of a network of these primary supplies by inducing downstream users to buy their components from local producers.

*Iron and Steel Industry*

1. To rationalize and reorganize the industry, especially in the production of billets, bars, wire rods, and light sections which face problems of inefficiency, high cost of production, and considerable overcapacity.
2. To undertake programmes towards increasing efficiency, upgrading technology, and reducing production cost so as to enable the industry to be competitive in the export market.

*Textile and Apparel Industry*

1. To undertake a comprehensive rationalization and modernization programme through improvement and upgrading of existing facilities with the support of adequate incentives, particularly in the textile sub-sectors of spinning, weaving, and knitting.
2. To increase productivity and technology absorptive capacity and enhancement of manpower training and technology transfer in order to achieve international competitiveness.

All the above development objectives indicate the importance of technological upgrading in these industries, especially in the light of the country's export-led growth. Given the importance of these industries in the country's future industrialization programme, their technology requirements and the impact that government policies will have on these requirements will form a critical component of industrial planning strategy. The principal thrust of the above objectives are closely intertwined with increasing productivity, competitiveness, and efficiency, which must ultimately be related to both technological enhancement and manpower upgrading. In this respect, it is thus critical that industrial strategies be closely linked not only to technology development policies but also to human resource development. An important issue here is whether the existing institutional framework for policy-making has the capacity to integrate the various components of industrial, S & T, and educational policies.

Although the IMP does not recommend specific measures for upgrading domestic technological capability, it nevertheless gives due emphasis to the critical role of such capability. The Plan, as implied above, reiterates the vital role of industrial advancement if future economic growth is to be sustained so that ultimately Malaysia will join the ranks of the newly industrialized countries by the mid-1990s. Indigenous technology development would be a crucial component of this process so that productivity and efficiency in all the industrial sectors and international competitiveness could be enhanced.

In terms of human resource development, for instance, the importation of highly sophisticated technologies has to be complemented by an equally rapid increase in skill intensity so that a higher proportion of industrial labour is equipped with the advanced technical skills required to control and operate complicated production processes. In this way, a manpower base will be created, enabling the absorption of more sophisticated technologies that will result in increased industrial activities. More importantly, there is a need to develop indigenous skills in product design and production technology if domestic industries are to progress independently of imported foreign technology and expertise. However, the IMP also recognizes that there is a lack of technological capability within domestic industries, and this creates a major impediment to achieving a higher level of industrialization. Thus, according to the IMP:

> The modern industries in Malaysia which demand high technology are either foreign owned or joint-ventures involving foreign equity or foreign technical collaboration and therefore have direct access to foreign technology from their principals. This complacency, probably derived from easy access to foreign partners, has hindered the formulation of any coherent and comprehensive policy designed to develop indigenous industrial technology in Malaysia (apart from the technology in the area of primary products). As a result, the policy instruments and institutional mechanism to foster industrial technology remains [*sic*] inadequate with little R & D activities. In addition, little attention is paid to the generation of a minimum level of indigenous technology which is necessary to absorb technology from foreign sources and adapt them to gain comparative advantage in the market. Though technology and its absorptive capability are the critical elements for industrialisation, little incentives are provided for fostering their development.

It is therefore important that technology development be actively promoted in order to build up a strong foundation for 'leap-frogging' towards a more advanced stage of industrialization. The implementation of the IMP is expected to increase manufacturing value-added, improve industrial skills, and create new export niches. Beyond the 1990s, the manufacturing sector will be looked upon not only as the major source of employment creation, but more importantly, as a source of increasing demand for skilled labour and the vehicle for enhancing domestic technological capability.

TABLE 3.8

Foreign Investment in Approved Projects by Industry, 1985–1989 ($ million)

| *Industry* | *1985* | *(%)* | *1986* | *(%)* | *1987* | *(%)* | *1988* | *(%)* | *1989* | *(%)* |
|---|---|---|---|---|---|---|---|---|---|---|
| Food manufacturing | 58.6 | 6.1 | 293.8 | 17.4 | 202.6 | 9.8 | 571.1 | 11.7 | 290.7 | 3.4 |
| Beverages and tobacco | 1.1 | 0.1 | 2.3 | 0.1 | 4.2 | 0.2 | 7.1 | 0.1 | – | – |
| Textiles | 31.3 | 3.3 | 31.4 | 1.9 | 55.4 | 2.7 | 238.8 | 4.9 | 511.2 | 6.0 |
| Leather products | 0.4 | – | – | – | – | – | 0.4 | – | 18.5 | 0.2 |
| Wood products | 11.6 | 1.2 | 12.6 | 0.7 | 120.5 | 5.8 | 198.8 | 4.1 | 980.0 | 11.4 |
| Furniture and fixtures | 6.0 | 0.6 | 1.1 | 0.1 | 2.1 | 0.1 | 72.3 | 1.5 | 129.7 | 1.5 |
| Paper and printing | 101.9 | 10.6 | 12.3 | 0.7 | 79.7 | 3.9 | 34.2 | 0.7 | 294.2 | 3.4 |
| Chemicals | 29.4 | 3.1 | 42.0 | 2.5 | 325.8 | 15.8 | 763.5 | 15.6 | 993.6 | 11.6 |
| Petroleum and coal | 0.8 | 0.1 | 876.3 | 51.9 | – | – | – | – | 216.0 | 2.5 |
| Rubber products | 29.8 | 3.1 | 71.1 | 4.2 | 191.3 | 9.3 | 662.7 | 13.6 | 360.7 | 4.2 |
| Plastic products | 19.1 | 2.0 | 92.3 | 5.5 | 104.3 | 5.1 | 272.3 | 5.6 | 215.2 | 2.5 |
| Non-metallic mineral products | 110.8 | 11.6 | 26.3 | 1.6 | 79.7 | 3.9 | 73.4 | 1.5 | 336.4 | 3.9 |
| Basic metal products | 148.1 | 15.4 | 25.3 | 1.5 | 82.3 | 4.0 | 612.7 | 12.6 | 439.4 | 5.1 |
| Fabricated metals | 43.8 | 4.6 | 19.6 | 1.2 | 8.6 | 0.4 | 147.4 | 3.0 | 500.6 | 5.8 |
| Machinery | 43.7 | 4.6 | 21.5 | 1.3 | 23.4 | 1.1 | 12.2 | 0.2 | 143.3 | 1.7 |
| Electrical and electronic products | 110.7 | 11.5 | 97.2 | 5.8 | 752.3 | 36.5 | 1,151.9 | 23.6 | 2,720.8 | 31.8 |
| Transport equipment | 186.4 | 19.4 | 53.7 | 3.2 | 12.6 | 0.6 | 22.4 | 0.5 | 136.8 | 1.6 |
| Scientific equipment | 5.7 | 0.6 | – | – | – | – | 13.9 | 0.3 | 209.9 | 2.4 |
| Miscellaneous | 20.1 | 2.1 | 9.1 | 0.5 | 15.2 | 0.7 | 23.0 | 0.5 | 70.9 | 0.8 |
| Total | 959.3 | 100.0 | 1,687.9 | 100.0 | 2,060.0 | 100.0 | 4,878.1 | 100.0 | 8,567.9 | 100.0 |

*Source*: MIDA (adapted from Low Peng Lum, 1990).

A significant aspect is the ever dominant role of direct foreign investment in manufacturing as indicated in the preceding sections. While the electrical and electronics industry has consistently enjoyed substantial foreign investments, the RBIs, such as the chemical and petroleum, wood-based, and rubber products industries, have begun to attract foreign interests, particularly since 1985. Other industries which have attracted increased foreign investment are the textile and textile products, metal products, and plastic products industries. With the exception of 1985 and 1986, the electrical and electronics industry became the focus of direct foreign investment flows during the second half of the 1980s. During the 1987–9 period, foreign investment in this industry not only accounted for the largest share during the three years (36.5 per cent of the total in 1987, 23.6 per cent in 1988, and 31.8 per cent in 1989) but also experienced rapid annual increases in the amount of investment: 53.1 per cent in 1988 and 136.2 per cent in 1989 (Table 3.8).

The rapid expansion of the industry is also related to the fact that 'Malaysia is now equipped towards greater automation and the production of higher value-added products in view of the well established infrastructure and the supporting industries' (Malaysia, 1989b: 56). Furthermore, in a MIDA Industrial Trends Survey on business expectations for the period July–December 1989, it was observed that 80 per cent of the firms in the industry, in order to keep pace with the rapid changes in the industry world-wide, would continue to invest further to upgrade their existing plants and machinery to gear themselves towards more automation and the manufacture of higher value-added products (Malaysia, 1989c: 49).

The chemical industry also occupied a prominent place in terms of foreign investment, accounting for 15.8 per cent of total foreign investment in 1987, 15.6 per cent in 1988, and 11.6 per cent in 1989. The rubber products industry has sustained substantial increases in investments since 1987, following the continued strong external demand, especially for rubber gloves. The increased foreign interests in some of these industries can be attributed to the need for establishing ancillary and supporting industries to service the large number of export-oriented electrical and electronics projects which have been established mostly in the FTZs.

# 4
# Global Trends: Implications for Malaysia

## The Pace of Internationalization

As stated in the IMP, the formulation of any future industrial strategy will have to take into account the rapidly changing conditions of the world economy. This changing environment, generally influenced by the ever increasing competition for market niches, is likely to be translated into economic adjustments at the macro and micro levels—mainly as espoused by the advanced industrial countries because of their capacity to do so—which will have repercussions on the development of all Third World countries. This will ultimately affect Malaysia's own long-term industrial growth.

In the context of the Asia–Pacific region, two principal changes since the mid-1980s seem to have had a tremendous impact on the economic performance of most countries: first, the increasing pace of internationalization of production and, secondly, the increasingly important role of the Asian newly industrialized countries in manufactured export trade. Since the early 1970s, it has become increasingly evident that the MNCs of the advanced industrial countries, in tandem with their respective government policies, have successfully 'internationalized' the development of offshore facilities. The establishment of FTZs or Export Processing Zones (EPZs) in many developing countries, including Malaysia, is testimony to this new phase of global relocation of industries from the industrial countries. The massive spread of these offshore facilities since the mid-1980s adds a new dimension to the relationship between industrial countries and developing countries, exhibiting a division of labour in the world economy through the imposition of an intra-firm division of labour within the MNCs.

The intra-firm transfers of inputs and outputs, as well as their export from offshore bases or EPZs, now constitute a significant proportion of the import–export trade of most industrial countries as well as many middle-income developing countries. For the MNCs, this internal division of labour strongly implies, particularly in manufacturing activities, that they still have control over technology and innovation, while the routine and standardized production and marketing of products for domestic markets are located in the developing countries.

The increasing integration of national economies and the increasingly complex systems of co-ordination and telecommunications among dispersed production units, together with standardized manufacturing techniques and a world-wide marketing system, have accentuated the dominance of the MNCs. This dominance allows them not only to widen their market shares, but more importantly to stifle indirectly the entry of newcomers through product differentiation. Their strength in R & D activities, driven by the desire to remain competitive and the need to be at the frontier of technology development, has consolidated their oligopolistic positions. The manner in which the MNCs dominate world trade tends to make export-led industrialization a difficult task for most of the developing countries without the participation of the MNCs, at least at the initial stages, either through equity participation or technology licensing.

There appears to be limited options for a small economy such as Malaysia to develop export-oriented industries in sectors such as electrical and electronic products, consumer durables, chemicals, transport equipment, and even textiles without the necessary links with the MNCs. There is also evidence to indicate that the latter are reluctant to part with their technologies which, from their own perspective, could affect their future control on technology development and markets and their practice of sourcing components from their parent or affiliate companies. This, of course, has an important bearing on the speed or pace at which Malaysia as a developing economy can develop its own technological capability.

## Japan and the Emergence of the Newly Industrialized Countries

A related development since the mid-1980s has been the relocation of Japanese small- and medium-scale enterprises in the other countries of the Asia–Pacific region. The availability of relatively cheap labour in these countries has been identified as a critical factor in this relocation (*New Straits Times*, 19 April 1990). In this respect, Malaysia as well as other ASEAN countries, particularly Thailand, have become locations from which such enterprises can re-export their products to Japan or supply larger Japanese companies in the host countries (Urata, 1989). If these companies expand more rapidly than the capacity of domestic industries to produce equivalent products, it may well hinder the development of domestic supporting or ancillary industries.

The emergence of the Asian newly industrialized countries (NICs)—South Korea, Taiwan, Hong Kong, and Singapore—as major exporters of more mature consumer products has significant repercussions on the relationship between the Asian NICs and developing countries. These

TABLE 4.1
Per Capita GDP Growth of Economic Groupings, 1973–1988 (percentage)

| | *1973–80* | *1980–5* | *1986* | *1987* | *1988* |
|---|---|---|---|---|---|
| East Asia | 4.6 | 6.4 | 5.8 | 6.8 | 9.3 |
| Sub-Saharan Africa | 0.5 | –3.7 | 0.8 | –4.4 | –0.2 |
| South Asia | 2.0 | 2.9 | 2.2 | 0.9 | 5.6 |
| Latin America and Caribbean | 2.5 | –2.2 | 1.8 | 1.9 | –0.9 |
| All developing countries | 2.7 | 1.2 | 2.7 | 2.5 | 3.5 |
| OECD countries | 2.1 | 1.7 | 2.1 | 2.7 | 3.3 |

*Source*: World Bank (1989).

repercussions include enhancement of trade between the NICs and the developing countries—particularly those in the Asian region—and the flow of investment from these NICs into these countries. The economic performance of these NICs is well reflected in Table 4.1, indicating that their growth was relatively rapid, outperforming all other country groupings. During the 1973–85 period, for instance, the annual per capita GDP growth of the East Asian countries was well above the average for developing countries as a whole; the divergence is particularly noticeable during the 1980–5 period. The difference in these growth rates continued during the next three years, with the East Asian countries achieving a growth rate of 9.3 per cent in 1988, and the developing countries achieving 3.5 per cent.

This phenomenon has become the focus of a new international division of labour, especially with the expansion of East Asia's own MNCs and their extremely successful export performance. What is perhaps more significant in this respect is the clear response of the East Asian countries to the rapidly changing economic and technological environment within which they have had to operate internationally. This has been made possible by clearly defined strategies which prioritize export-led growth induced by new technologies in product and process development.

Looking at country performance during the 1980s, growth in the Asian NICs tended to outstrip that of ASEAN, but in the latter part of the decade the divergence lessened, with some moderation in the economic performance of the Asian NICs, and with the ASEAN economies, with the exception of the Philippines, consolidating their recovery from the mid-1980s downturn (Table 4.2). The economic performance of each of these countries is also reflected in its current account status as exemplified in Table 4.3. The table indicates that, in 1987, the major Asian surplus economies were Japan, Taiwan, and South Korea, with surpluses of US$87.0 billion, US$29.8 billion, and US$9.9 billion respectively.

TABLE 4.2
GDP Growth Rates in ASEAN and the Asian NICs, 1965–1987

| | *GDP Annual Growth Rate* | |
|---|---|---|
| *Country* | *1965–80* | *1980–7* |
| Indonesia | 8.0 | 3.6 |
| Malaysia | 7.4 | 4.5 |
| Philippines | 5.9 | –0.5 |
| Singapore | 10.1 | 5.4 |
| Thailand | 7.2 | 5.6 |
| Hong Kong | 8.6 | 5.8 |
| South Korea | 9.5 | 8.6 |

*Source*: World Bank (1989).

The rapid growth of foreign investment by South Korean firms since 1986 has, in the main, been attributed to the turn around in the country's current account that has expanded substantially since then. A significant effect arising from this situation was the relaxation of government policy towards South Korean investments overseas. In fact, the South Korean government even promoted certain types of investments overseas by providing loans and consultancy services, especially in declining domestic industries and RBIs (Park, Eul-Yong, 1989). On the other hand, the major current account deficit countries in Asia are India, Indonesia, Pakistan, and the Philippines. Thus, one of the challenges of the 1990s is the efficient recycling of such surpluses. In this sense, Japan could provide the lead with the Asian NICs close behind and ASEAN countries following the bandwagon, with each country within ASEAN trying hard to attract new industries or investments from both Japan and the Asian NICs.

How these relationships can be strengthened will critically depend on the responses within each country and the speed with which each country succeeds in adjusting to the new environment (Lo et al., 1987: 29–99). A potential area which could be exploited is Japan's future role as an increasingly important export market for the manufactured products of other countries in the region, although this will also depend on the success of structural adjustments within Japan that would speed up its domestic demand expansion instead of relying on its traditional export-led growth (Takenaka, 1989). Since the high appreciation of the yen, Japanese manufacturing enterprises have successfully diversified their products, expanded into other lines of business, and significantly increased their overseas investments. Japanese industries in general have thus adjusted appropriately to the strong yen through technological adaptation and innovation, with the full support of management and labour, which consequently allows them to become more competitive in the international markets (Park, Yung-Chul, 1989).

TABLE 4.3
Current Account Balance of Asian Developing Economies, 1987
(US$ billion)

| | *Exports* | *Imports* | *Trade Balance* | *Current Account Balance* |
|---|---|---|---|---|
| World | 2,279.10 | 2,251.00 | 28.10 | −43.20 |
| USA | 249.57 | 409.85 | −160.28 | −153.95 |
| EC | 916.81 | 886.16 | 30.64 | 39.14 |
| Japan | 224.63 | 128.17 | 96.46 | 87.00 |
| Asian NICs | 175.22 | 149.31 | 25.91 | 29.76 |
| Taiwan | 53.22 | 32.44 | 20.78 | 18.17 |
| South Korea | 46.24 | 38.59 | 7.66 | 9.85 |
| Singapore | 27.28 | 29.82 | −2.54 | 0.54 |
| Hong Kong | 48.48 | 48.46 | 0.01 | 1.20 |
| South-East Asia | 52.23 | 43.29 | 8.94 | −0.44 |
| Indonesia | 17.21 | 12.71 | 4.50 | −2.15 |
| Philippines | 5.72 | 6.74 | −1.02 | −0.50 |
| Thailand | 11.60 | 12.02 | −0.42 | −0.37 |
| Malaysia | 17.71 | 11.82 | 5.89 | 2.57 |
| South Asia | 18.79 | 29.36 | −10.57 | −6.85 |
| Bangladesh | 1.08 | 2.45 | −1.37 | −0.34 |
| India | 11.88 | 17.66 | −5.78 | −5.19 |
| Myanmar (1986) | 0.33 | 0.62 | −0.29 | −0.29 |
| Nepal | 0.16 | 0.51 | −0.35 | −0.12 |
| Pakistan | 3.94 | 6.25 | −2.32 | −0.56 |
| Sri Lanka | 1.39 | 1.87 | −0.47 | −0.34 |
| China | 34.73 | 36.40 | −1.66 | 0.30 |

*Source*: Quoted from Chee (1990).

However, all the above changes at the country or industry level will inevitably force each economy to make the necessary economic adjustments, which will manifest themselves through the changing dynamic comparative advantage of the different economies according to the different stages of their industrialization process. In the Asia–Pacific regional context, China, for instance, is at the early stage of its industrial development, but undergoing rapid sectoral transformation and adjustments in its economic structures. Such changes will undoubtedly influence the pace of growth of the region in the 1990s. ASEAN countries (with the exception of Singapore), on the other hand, finding themselves in between the early industry and middle industry stages, are expected to undergo rapid structural changes as they move into the late industry stage, while the four Asian NICs have been at the late industry stage since the mid-1980s and are faced with the prospect of developing high-tech technologies, including microchips, laser technology, robotics,

TABLE 4.4
Foreign Investment in Approved Projects by Major Countries, 1985–1989 ($ million)

| *Country* | *1985* | *1986* | *1987* | *1988* | *1989* |
|---|---|---|---|---|---|
| Japan | 264.4 | 116.3 | 715.1 | 1,222.0 | 2,682.2 |
| | (27.6) | (6.9) | (34.7) | (25.1) | (31.3) |
| Taiwan | 31.9 | 10.8 | 243.0 | 829.6 | 2,119.4 |
| | (3.3) | (0.6) | (11.8) | (17.0) | (24.7) |
| Singapore | 100.2 | 183.7 | 258.5 | 419.6 | 910.9 |
| | (10.5) | (10.9) | (12.5) | (8.6) | (10.6) |
| United States | 111.9 | 53.6 | 162.7 | 535.2 | 320.8 |
| | (11.7) | (3.2) | (7.9) | (11.0) | (3.7) |
| United Kingdom | 26.9 | 49.5 | 76.8 | 196.5 | 764.1 |
| | (2.8) | (2.9) | (3.7) | (4.0) | (8.9) |
| Hong Kong | 28.2 | 55.9 | 88.9 | 298.4 | 352.1 |
| | (2.9) | (3.3) | (4.3) | (6.1) | (4.1) |
| South Korea | 25.0 | 4.0 | 3.6 | 41.8 | 189.0 |
| | (2.6) | (0.2) | (0.2) | (0.9) | (2.2) |

*Source*: MIDA (adapted from Low Peng Lum, 1990).
*Note*: Figures in parentheses indicate percentages of total foreign investment in manufacturing in that particular year.
Total Foreign Investment:
1985—$959.3 million
1986—$1,687.9 million
1987—$2,060.0 million
1988—$4,878.0 million
1989—$8,568.1 million

genetic engineering, new industrial materials, etc. (Lo and Song, 1987: 463–95).

As they go through the various stages of industrialization, ASEAN countries (such as Malaysia), China, and the Asian NICs will increasingly rely on Japan as the major source of technology, and for many even capital, for their industrialization to proceed further. In the case of Malaysia, its dependence on Japanese investment can be clearly gauged from Table 4.4. During the 1985–9 period (with the exception of 1986), Japan's share in domestic manufacturing was between 25 and 35 per cent of total foreign investment, while Taiwanese investment has been playing an increasingly significant role since 1987. In the context of Japanese direct investment in the Asian countries, however, Indonesia seems to be a favoured location both with respect to its total stock of investment as well as manufacturing investment as indicated in Table 4.5. Although Indonesia's share has declined relatively during the 1978–88 period, Japanese total investment in Indonesia was still substantial at US$9,804 million in 1988, which accounted for 30.8 per cent of the total Japanese investment in the Asian region for that year, while the respective figures for manufacturing investment were US$2,955 million and 24.3 per cent.

TABLE 4.5
Stock of Japanese Direct Investment in Major Asian Countries, 1978, 1983, and 1988 (US$ million)

| | *Total Investments* | | | | | | *Manufacturing Investments* | | | | | |
|---|---|---|---|---|---|---|---|---|---|---|---|---|
| *Country* | *1978* | *(%)* | *1983* | *(%)* | *1988* | *(%)* | *1978* | *(%)* | *1983* | *(%)* | *1988* | *(%)* |
| Hong Kong | 715 | 9.5 | 1,825 | 12.7 | 6,167 | 19.4 | 145 | 4.3 | 215 | 3.8 | 492 | 4.0 |
| South Korea | 1,007 | 13.4 | 1,312 | 9.1 | 3,248 | 10.2 | 697 | 20.8 | 839 | 14.6 | 1,589 | 13.1 |
| Taiwan | 284 | 3.8 | 479 | 3.3 | 1,791 | 5.6 | 266 | 7.9 | 439 | 7.7 | 1,473 | 12.1 |
| Indonesia | 3,745 | 49.9 | 7,268 | 50.7 | 9,804 | 30.8 | 1,166 | 34.7 | 2,001 | 34.9 | 2,955 | 24.3 |
| Malaysia | 471 | 6.3 | 764 | 5.3 | 1,833 | 5.8 | 302 | 9.0 | 533 | 9.3 | 1,350 | 11.1 |
| Philippines | 434 | 5.8 | 721 | 5.0 | 1,120 | 3.5 | 152 | 4.5 | 290 | 5.1 | 510 | 4.2 |
| Singapore | 541 | 7.2 | 1,383 | 9.6 | 3,812 | 12.0 | 395 | 11.8 | 1,009 | 17.6 | 1,990 | 16.4 |
| Thailand | 309 | 4.1 | 521 | 3.6 | 1,992 | 6.3 | 233 | 6.9 | 390 | 6.8 | 1,456 | 12.0 |
| China | – | – | 73 | 0.5 | 2,036 | 6.4 | – | – | 11 | 0.2 | 349 | 2.8 |
| Total | 7,506 | 100.0 | 14,346 | 100.0 | 31,803 | 100.0 | 3,356 | 100.0 | 5,727 | 100.0 | 12,164 | 100.0 |

*Source*: Ministry of Finance, Japan (quoted in Urata, 1989).

While Hong Kong is an important recipient of Japanese total investment, accounting for 19.4 per cent in 1988, its share of Japanese manufacturing investment is relatively small. As shown in the table, South Korea, Singapore, Thailand, and Taiwan are also important locations for Japanese investment, with Malaysia trailing behind, although its share, especially in manufacturing investment, has become increasingly significant. China, too, has been an important recipient of Japanese total direct investment, particularly in 1988 when its share was 6.4 per cent, although its share of Japanese manufacturing investment is still very small in comparison to the other Asian countries.

Given their innovative edge, however, the industrial countries such as Japan will invest in new industrial products utilizing new or emergent technologies and their already well-developed industrial manpower. To a lesser extent, this may also apply to the other Asian NICs, particularly South Korea and Taiwan. This industrial restructuring and shift in the position of the different economies within the context of global specialization and division of labour are not necessarily smooth. For the industrial countries, this will mean the rationalization and modernization of industries as well as the re-equipping and retraining of their skilled manpower. For the developing countries, the challenges are certainly much more demanding, requiring continuous assessment of existing policies covering a wide range of areas including human resource development, science and technology policies, and industrial relocation.

In fact, a new pattern of subregional division of labour has emerged in the Asia–Pacific region. The ASEAN countries and China are catching up with the NICs in labour-intensive industries while the NICs are catching up with Japan in technology- and knowledge-intensive manufactures. This has often been referred to as the 'flying geese' pattern. Suh (1989: 107–30), for instance, argues that the countries in the region are becoming more interdependent because of complementary relations in production inputs, technologies, and markets. Particularly during the 1970s, the industrial growth of these countries was principally derived from their competitive advantage in light manufacturing exports geared to meet increasing demands in the industrial countries. However, in the industrialization process, these countries gradually shifted to intermediate and capital goods industries, hence broadening their industrial bases and enhancing their industrial structures.

The 'flying geese' pattern is also discernible in heavy industries whose production focus seems to have shifted from the United States to Japan in the 1970s, paralleling a shift in the light manufactured goods industry from Japan to the Asian NICs. The relative performance of Japan, the Asian NICs, and ASEAN in trade in the three categories of manufactured goods—non-durable consumer goods, durable consumer goods, and capital goods—over the 1970s and 1980s is reflected in the pattern of trade specialization that has emerged amongst the three economic entities. Japan has become a net importer of non-durable consumer goods since the early 1970s, while the Asian NICs reached a peak as net exporters of this product category in the late 1970s. Some ASEAN countries became net exporters of consumer non-durables in the early 1970s, and others will most likely catch up with the NICs in this category in the 1990s.

In the durable consumer goods category, however, ASEAN remains a net importer and is likely to remain so until the mid-1990s. The Asian NICs became net exporters in this category in the early 1970s and look set to become potential competitors to Japan in the early 1990s. However, it is clear that Japan will remain dominant as a net exporter of capital goods for some time to come, although the Asian NICs will probably reach self-sufficiency in the 1990s and become net exporters themselves as they already are in some areas. ASEAN will continue to depend on imports of capital goods from Japan, other industrial countries, and the Asian NICs during the 1990s. The Asian NICs will, in fact, diversify from the processing of products into design, R & D, testing, and marketing and will increasingly become important exporters of technology which includes complete plants, capital goods, and consultancy and training services (Chen and Wong, 1989: 204–39).

The role of MNCs, including those from Japan and the Asian NICs, especially South Korea, in trade and investments in the region have strengthened further the globalization of production in the Asia–Pacific region. South Korean MNCs, for example, have been credited with their ability to assimilate technology from the industrial countries, adapt it for domestic conditions as well as re-export the refined technology to other developing countries which face the same resource conditions or to again readapt it to suit the latter's particular needs (United Nations, 1985: 71–2). The importance of technology, finance, and market channels has been increasingly demonstrated by the activities of the MNCs. It remains for the NICs to ensure that effective transfer of technology does take place as well as the enhancement of the spin-off effects of direct foreign investment in the host countries; this is the difficult part.

Since the early 1980s, trends in direct foreign investment flows have shown a distinct development towards greater structural interdependency in the world economy, and in particular in the Asia–Pacific region which has fast emerged as the most dynamic part of the world economy since the late 1980s. At the same time, enhanced intra-firm division of labour and the possibility of new decentralized forms of organization of production is the likely future direction of MNC activity in the region. In this global and regional economic context, the thrust for Malaysia in the future would thus have to be in finding a niche in the world manufacturing export market and expanding the industrial base through the enhancement of technology development, so that in the longer term the interdependency will become more beneficial to domestic manufacturing enterprises.

## Sustained Industrial Growth and New Technologies

An economy with a resilient industrial structure and a well-endowed technological base will be assured of long-term sustained industrial growth. This structure must also be flexible enough to respond to the changing external environment and opportunities; this flexibility requires a strong science and technology capability. The traditional comparative advantage that Malaysia enjoys in terms of its primary products and labour-intensive industries is quickly changing, not only because of the emergence of new players in the global trading system but also the emergence of new technologies and the rapidity of responses that these entail. It is the area of new technologies that will have a significant impact on the future thrust of economic growth. The momentum for such growth must then be sustained by the capacity to acquire and assimilate new technologies in domestic industries, and the very sustenance of these industries will depend on the rate at which technological change and innovation take place within the manufacturing sector.

During the 1970s and 1980s, perceptions regarding technological advancement have changed substantially as a consequence of the emergence of Japan and, to a lesser extent, the Asian NICs, especially South Korea, as nations which have achieved rapid economic growth mainly through the effective management and utilization of technology as the principal catalyst for economic advancement. For these countries, technology has become a critical factor for national economic expansion given their relatively scarce natural resources. Citing the case of Japan during the 1970s, Sakaiya (1991: 16–17) has this to say:

> As much as petroleum-driven growth may have been a boon, the beneficial effects it provided were not distributed equally among the nations of the world. The countries that imported raw materials and agricultural products got more out of the boom than those that exported them; the importers, of course, were the countries that had advanced industrial economies. They could use their existing facilities and human resources to exploit the cheap and abundant raw materials provided by the less advanced nations. No people derived greater advantage from this situation than the historically resource-poor Japanese.

New technologies, reflecting the tremendous advances in industrial innovations during the 1980s, principally emanating from the industrial countries, will have a profound impact on the global industrial environment. These are found in areas such as micro-electronics and communications, biotechnology, automated manufacturing technology, and new materials technology. The development and application of new technologies is probably the most active area of science and technology policy in the industrial countries. Their technologies are indeed expanding very rapidly as they have the capacity to mobilize their human and capital resources, and it is this capacity that will fundamentally ensure their future competitiveness and economic well-being. Even countries where the state traditionally took a non-interventionist posture could not maintain this position and have now become active in assisting and managing national programmes in key and emerging technologies.

At the same time, collaboration between government and industry within industrial countries in advancing and applying these technologies is increasingly nurtured. A fundamental principle in this context is, of course, collaboration and the need for enhanced public sector–industry interaction. The extent of such co-operation and the numerous institutional or administrative approaches that have been established to maximize resource utilization in planning, financing, and management is unprecedented in several industrial countries. This development appears to be motivated by the ever increasing costs and risks of R & D and innovation as well as the joint concern for improving industrial competitiveness at the level of the firm. Co-operation is not restricted to government and industry, but universities and research institutes engaged in basic research are also increasingly involved in national R & D efforts associated with new technologies.

These developments have been driven, to a large extent, by technological rivalry among nations and industrial enterprises, especially among the MNCs, arising from intense competition for world markets. The wide application and dissemination of these new technologies in these countries will have important repercussions on the basic structure of industrial production in terms of production costs, labour utilization, productivity, and industrial relocation, thereby influencing the future shape of the international system of comparative advantage. For instance, the progress in new electronic-based technologies during the 1980s has made possible the adoption of more computer-integrated designing and manufacturing systems. While this permits new products to be commercialized very quickly once they have been designed, that is, shortening the lead-time from the product conception stage to the design stage, it also allows manufacturers to respond rapidly and flexibly to customers' specifications and to changing market tastes. Consequently, there is a perceptible trend for manufacturing enterprises to emphasize product diversification, design, distribution, and servicing of their products, apart from production itself.

The rapidity of technological change, the increasing number of suppliers, faster technology diffusion, and increased automation have together tended to shorten the life-span of new products, resulting in product obsolescence becoming a major risk factor. There is thus constant pressure on manufacturing enterprises to maximize returns from each new product as quickly as possible before such obsolescence sets in. All these changes imply that Malaysia, being a low-cost producer relying principally on relatively cheap labour, has no other option but to adapt as quickly as possible so as to upgrade its technological capability and skilled manpower. This is particularly relevant given that export-led growth will become the focus of the country's industrialization strategy in the future. In many industrial sectors, competitive advantage in the 1990s now depends not only on lower production costs, but more importantly on speedy and reliable delivery, high quality, and the ability to expand one's range of products and services to fit customers' changing needs, especially in the international market. At the same time, the need for maintaining high-quality products implies that manufacturers must enhance their level of automation. Consequently, the role of labour in the process of production has to change. A country with an abundance of relatively low-cost labour can no longer rely on such labour to maintain its international competitiveness.

All these changes suggest that Malaysia must also equip itself with the necessary industrial adjustments as envisaged by the IMP. Policies towards this end must include a comprehensive rationalization programme for declining industries, as well as a modernization programme for existing industries, including the small- and medium-scale industries. In the long run, however, these adjustment policies must be complemented by strategies that would strengthen technological capability within the domestic industries.

## Global Adjustments and International Technology Diffusion

The need for rapid technological diffusion within the economy has become more critical in the light of global macroeconomic adjustments and technological changes and breakthroughs, especially since the early 1980s. These may be seen in terms of macroeconomic policies, whether emanating from the more advanced industrial countries either individually or as economic blocs, affecting international trade as well as technology flows. The continuing restructuring in the industrial countries and the prevailing mood of protectionism add to instability and uncertainty in the world economy. The current trends towards bilateralism and regionalism, particularly among the major industrial countries, may well produce an adverse effect on the open multilateral character of the global trading system. For example, apart from near total economic integration within the European Community (EC) by the end of 1992, the European Community and the European Free Trade Association (EFTA) have agreed to set up a European Economic Area (EEA) in 1993, forming the world's largest free trade area with 380 million consumers and representing 46 per cent of world trade (*The Economist*, 26 October–1 November 1991: 77–8).

At the same time, an inconclusive end to the Uruguay Round of Multilateral Trade Negotiations may further increase protectionism, and hence retard growth in international trade. Such complexities become all the more disconcerting for developing economies given the rapidity in the development of new emerging technologies in areas such as microelectronics, biotechnology, and advanced materials which are primarily driven by a process of intensifying competition and shifting competitive advantages among the MNCs of Japan, the United States, and the EC (Commonwealth Secretariat, 1985: 32–42).

Since the 1980s, industrial R & D spending has been undoubtedly concentrated in the large MNCs, which account for approximately 75 per cent of all industrial R & D funding in the industrial countries (OECD, 1985: 63). MNCs such as Siemens, Hoechst, IBM, Hitachi, Toshiba, and Philips obviously exert a commanding influence on industrial R & D and technological innovation in the world. These giant corporations set the pace and direction of R & D and through their domestic and foreign affiliates should, if they wish, be able to become primary agents for the international transfer of technology.

In most of the advanced countries, industrial expenditures for R & D continue to increase at a rapid pace, and, interestingly enough, they continue to increase even during periods of economic recession. There are several possible explanations for the continuing growth of R & D spending (OECD, 1985: 61). First, manufacturing firms generally regard R & D as a high yield investment, particularly in a situation where there is rapid technological innovation and promising prospects for advances and applications. The recession itself may have contributed by intensifying competition and leading firms to rely increasingly on their R & D as a source of competitiveness. Secondly, government policies for promoting industrial R & D may well have stimulated additional spending. Thirdly, to some extent the introduction of new technologies is also a response to the growing market penetration of imports from the NICs.

In the midst of these global economic adjustments, the Asia–Pacific economies, centred around Japan and the four Asian NICs, represented a focal point of relatively high growth and economic ascendancy during the 1970s and 1980s. Japan provided the lead in this remarkable performance in spite of the appreciating yen, followed closely by the NICs in generating substantial trade surpluses, particularly with the United States. The relatively resource-rich countries of ASEAN, with the exception of Singapore, readjusting to the collapse of commodity prices in the early 1980s, have shown signs of economic recovery since the mid-1980s and, despite being 'latecomers' in the industrialization process, are pushing further into their second-phase industrial programmes, giving greater emphasis to export-oriented and higher-technology industries.

For some, the prospects for the 1990s remain equally optimistic. Thus, according to the Sixth Malaysia Plan (1991–5):

> ... the economies of the East Asia region will continue to remain strong, particularly the newly industrialised economies (NIEs) which have shown their ability to adjust successfully to adverse changes in the world economic environment. ASEAN will continue to benefit from inflows of direct foreign investment in the medium term. The export-led growth strategies adopted by these countries have helped to liberalise their economies and increase their internal strength to face adverse international economic development (Malaysia, 1991b: 16–17).

For a number of developing countries, such as Malaysia, if there is already an established entrepreneurial class and it is supported by the state through fiscal incentives and infrastructure, including S & T, there is a potential advantage in being a 'latecomer'. The option to acquire or purchase technologies already developed and proven in the industrial countries is available, and in that process, the development costs of such technologies are minimized while allowing domestic industries to achieve high levels of productivity and efficiency. While this may appear to be easy and straightforward, however, the pace of technological innovation is more rapid today than at any previous time, such that new technological frontiers are quickly shrinking. As mentioned earlier, product life cycles are being strikingly shortened, and developing countries like Malaysia would have to have a considerable capacity to enhance their learning curves to narrow the technological gap existing between themselves and the industrial countries. Even a strategy of 'creative imitation' or 'reverse engineering' (in which domestic manufacturers purchase particular products, take them apart, and reconstruct the same or similar products) requires very substantial investments, considerable flexibility, and the ability to respond immediately to technological changes. Therefore, a critical component of technology development is related to a country's capability to master its human, capital or financial, and other resources needed to maintain continued access to new technologies. To have access to the latter may, in many instances, also require the import of foreign capital into the domestic manufacturing sector as has been the experience of many countries.

While it is crucial to maintain a conducive environment to attract investments into the manufacturing sector, its growth momentum will also depend upon its ability to gain access to and garner effectively the potential of new emerging technologies, both to upgrade traditional industries and to seize new market opportunities generated by these technologies. It is in this sense that domestic enterprises and entrepreneurs must contribute effectively. When an export-oriented economy does not increase its market shares in the new growth industries but instead loses market shares because of declining competitiveness in its traditional industries, the process of industrialization becomes more difficult or may even stagnate. Unlike most of the developing economies, the Asian NICs have during the 1970s and 1980s upgraded their human resource and technology base such that their industries have substantially increased their value-added component to the stage where they are exceptionally competitive in the international markets.

Since the mid-1980s, production techniques at the firm level have been fundamentally altered by the world-wide diffusion of new microelectronics technologies as well as automation, as a result of which there has been a substantial reduction in the unskilled or semi-skilled labour component of many industries and product lines. In a sense, these trends will weaken those economies whose competitive advantages are largely based on labour costs. Such a phenomenon has clearly been the experience of Malaysia since the mid-1980s. As a consequence of these kinds of advancements, production facilities such as electronics assembly and garment manufacturing, which have been considered as the principal catalysts to labour-intensive operations in this country, have now become highly capital- and skilled labour-intensive with sophisticated equipment, including robots, gradually substituting labour (Fong, 1990: 173–207).

As global competitiveness will increasingly be dependent on a country's supply of highly skilled manpower, countries which are already well endowed with such manpower will have an upper hand relative to those with a largely unskilled or semi-skilled labour force. While the latter category of labour is abundant in the developing countries, the industrial countries on the other hand have not only an ample supply of highly skilled manpower to enhance their technological advancement but also, more importantly, the capacity to produce manpower that is well equipped to master or reorientate itself to new developments in technology. Thus, potential new entrants, especially the developing countries, wishing to export their manufactures will have to face increasing barriers.

Direct foreign investment has certainly been an important vehicle for the international diffusion of commercial technology. While most of these technology flows remain confined to the industrial countries, some of them end up in the developing countries, particularly those with the capacity to acquire and assimilate such technologies. Given that there are high barriers to the entry of new competitors in terms of international technology transfer and that the production of technology is costly, risky, and involves large-scale operations, only the large MNCs can invest substantially in R & D, especially in relatively advanced and sophisticated technology areas (United Nations, 1984: 5).

MNCs, being major players in direct foreign investment flows, are also subjected to numerous pressures to increase their technology exports, for at least three reasons. First, they need to extend the life cycle of mature technologies by means of global strategies or 'planned obsolescence'. Secondly, they are consistently searching for ways and means to recuperate the enormous costs incurred in R & D and in design and product development, for instance, through highly sophisticated transfer pricing techniques. Thirdly, they perceive that technology exports would allow them to penetrate markets which have hitherto been closed to their products. As implied earlier, the international diffusion of technology, nevertheless, has been selective and unequally distributed. This has been due primarily to the lack of finance, the limited capacity of the developing countries to monitor, unpackage, and absorb imported technologies, and to the tremendous economic and social costs involved in the transfer of technologies.

At the same time, fundamental changes have taken place in international financing or investment patterns since the mid-1970s, and these have significant implications for international technology diffusion. For instance, during the 1980s there were important interregional differences in the patterns and trends in external financing flows. Table 4.6 shows that the experience of middle-income East Asian economies differed from that of other developing regions. First, private sources of long-term lending are much more significant, accounting for 73 per cent of all long-term lending in 1987. Official development assistance (ODA) was a relatively insignificant part of capital flows to these East Asian countries, with the absolute level of development aid to the region being the lowest among the developing country groups. Secondly, the Asian region was the only one to record an increase in the flow of direct foreign investment between 1981 and 1987. There has also been a decline,

TABLE 4.6
External Financial Flows, 1981 and 1987

| | *Official Development Assistance (US$ billion)* | | *Private Long-term Lending as Percentage of Total* | | *Direct Foreign Investment (US$ billion)* | |
|---|---|---|---|---|---|---|
| *Country Group* | *1981* | *1987* | *1981* | *1987* | *1981* | *1987* |
| All developing countries | 24.5 | 30.4 | 74 | 56 | 10.2 | 9.5 |
| Sub-Saharan Africa | 7.1 | 11.1 | 58 | 32 | 1.3 | 0.8 |
| Middle-income East Asia | 1.8 | 2.2 | 75 | 73 | 1.8 | 2.5 |
| Low-income Asia | 5.3 | 7.2 | 45 | 48 | 0.2 | 1.3 |
| Latin America and Caribbean | 2.1 | 3.6 | 88 | 49 | 5.6 | 4.1 |

*Source*: World Bank (1989).

later followed by a severe reduction, in the flow of direct foreign investment towards developing countries, with two exceptions: China and the Asian NICs.

The developing countries' share of investment inflows declined from 27 per cent of the total during the 1981–3 period to 21 per cent during the 1984–7 period, while industrial countries, particularly the United States and the EC, increased their share of total inflows. Only in 1987 did investment flows to developing countries pass the level attained in 1981. This can be interpreted as a phenomenon in which rapidly growing developing countries with expanding internal markets, such as China and other Asian economies, continued to attract increased direct foreign investment (Chee, 1990). However, the shift in investment flows to the industrial countries may also be attributed to protectionism, the emergence of regional trading blocs (such as the impending single market of the EC at the end of 1992 and the United States–Canada Free Trade arrangement), and the growth of direct foreign investment in services, especially in financial services.

The flow of direct foreign investment within the industrial countries has thus undergone a period of radical restructuring. As Western European and, later, Japanese MNCs have expanded their investment in the United States, the latter's economy has become as much a 'host' as a 'home' country for MNCs. Japan has emerged as a major 'home' country for MNCs and a large source of foreign investment directed towards the United States and the EC, in the first place, and, secondly, towards the Asian NICs and the ASEAN region.

As shown in Table 4.7, the stock of Japanese direct investment overseas increased at a substantial rate during the 1978–88 period in terms of both total investment and the share of the total in manufacturing investment. Conversely, it is observed that the share going to Asian

TABLE 4.7
Stock of Japanese Direct Foreign Investment, 1978, 1983, and 1988
(US$ million)

| | *1978* | *1983* | *1988* |
|---|---|---|---|
| World total investments[a] | 26,809 | 53,131 | 186,356 |
| Developing regions | 15,150 | 28,390 | 71,786 |
| Percentage of world total | (56.5) | (53.4) | (38.4) |
| Asian developing countries | 7,668 | 14,552 | 32,227 |
| Percentage of world total | (28.6) | (27.4) | (17.3) |
| Manufacturing investments[a] | 9,174 | 16,952 | 49,843 |
| Developing regions | 6,619 | 10,536 | 19,307 |
| Percentage of manufacturing investments | (72.1) | (62.2) | (38.7) |
| Asian developing countries | 3,410 | 5,800 | 12,371 |
| Percentage of manufacturing investments | (37.2) | (34.2) | (24.8) |
| Manufacturing investments as percentage of total | (34.2) | (31.9) | (26.7) |

*Source*: Ministry of Finance, Japan (quoted in Urata, 1989).
[a]Excluding North America, Europe, and Oceania.

developing countries decreased over the years, from 37.2 per cent in 1978 to 24.8 in 1988 of manufacturing investment. Likewise, the share of Japanese total direct investment in developing countries as a whole declined dramatically, especially in 1988 when the share for developing countries was 38.4 per cent compared to 56.5 per cent in 1978 and 53.4 per cent in 1983. A similar situation was observed in the case of manufacturing investment, when the share that went to developing countries declined from 62.2 per cent in 1983 to 38.7 per cent in 1988.

These trends suggest that Japanese investment is increasingly located in the industrial countries. Even investments from the Asian NICs, such as South Korea, seem to follow similar trends. During the 1981–8 period, for instance, a total of US$537.8 million was invested by South Korean firms in the United States and Canada, accounting for 38.2 per cent of total South Korean investments during the period. These countries thus received the largest share of South Korean overseas investments followed by Asian countries with 23.2 per cent of the total and the Middle East with 17.0 per cent (Park, Eul-Yong, 1989). These investments, particularly in North America, were made not only because of the need to expand South Korea's market shares in the industrial countries but also because of the need to circumvent protectionism in these countries.

While ample empirical evidence of what this implies for international technology diffusion is still lacking, it is useful to note that with the possible exception of the Asian NICs, ASEAN, and China, the dissemination of technologies to developing countries has declined in relative terms. At the general level, technology sources have been increasing at a significant rate during the 1970s and 1980s, which, in theory, could enhance the scope for developing countries' technology sourcing, especially for those which are still at the early stage of their industrialization programme. However, on closer examination two fundamental yet conflicting trends appear. First, among the advanced industrial countries, a certain degree of convergence of technological capabilities already exists as reflected in the relative decline of US technological supremacy relative to Japan and the advanced countries of the EC and, secondly, for the developing countries, the potential for the sourcing of technology has drastically increased (Ozaki, 1991: 59).

The Asian NICs and some of the countries with potentially large domestic markets, especially Brazil, India, and China, have consolidated or even slightly enhanced their position relative to the leading industrial countries in certain industrial sectors. For the great majority of developing countries, however, the gap in terms of accessibility and the spread or diffusion of new generic technologies has drastically increased, thus significantly inhibiting their growth potential and international competitiveness.

In short, while technology diffusion has continued to expand globally, it has become even more skewed than before. The benefits to be derived from such diffusion certainly differ from country to country, depending to a large extent on a country's industrial and technological sophistication. An increasing number of developing countries are at best marginally integrated into the international system of technology flows. But the need for rapid technological changes has become more crucial in the light of global adjustments, especially since the early 1980s. Since then, new innovations in micro-electronics and information technology, robotics, biotechnology, and new materials have gradually entered into industrial production processes and are beginning to affect relative factor costs and the comparative advantage of production among the NICs and the rest of the world. Inter-industry integration of industrial production has increased through globalization of production, mainly through MNCs which are not only the major source of technology transfer but which also provide access to foreign markets and international financing.

While R & D in biotechnology, micro-electronics, and advanced materials will gain prominence during the 1990s, successful industrialization will increasingly depend on the utilization of the country's skills to develop and enhance new technologies. In Taiwan, for example, a plan has already been initiated to stimulate the entry of its manufacturing enterprises into the burgeoning biotechnology market. Although other countries in the Asian region have earmarked biotechnology as an industry with substantial potential, Taiwan has several advantages which enable it to be one of the early leaders in the industry. This includes its science and technology-oriented education system that caters for biotechnology developments, and the presence of many of its scientists in the United States working in this area. Furthermore, the government provides R & D funding through the National Science Council to stimulate the industry through the Development Centre for Biotechnology which has been established with the objective, among others, of developing products for private companies under contract, investing as a minority shareholder in promising new ventures, and initiating technology transfers from foreign companies to domestic firms (Shapiro, 1986: 58–9).

For Taiwan and many other developing economies, there is little doubt that new industries will emerge, but the most important influence on their development will depend on how new technologies are utilized to modernize existing industries. Since substantial benefits could be realized if the diffusion process were to be speeded up, it is imperative that such a process be facilitated within domestic industries. Unfortunately, in the case of Malaysia, the country's technological base since the 1980s has been unevenly developed amongst industries. As such, there is a critical need to be less dependent upon imported technologies while ways and means must be found to improve the capability of domestic industries to acquire the appropriate type of technologies, to adapt them to suit local conditions, and in the long run to innovate in existing capabilities.

Despite the constraints faced by Malaysia in terms of its technological capabilities, its export-led growth strategy must be complemented by measures to enhance product standards and quality in the domestic industries. Standard and quality control systems will play an increasingly significant role in enhancing the country's export capability; in this respect, manufacturing enterprises which are export-oriented must be encouraged to establish quality control measures to ensure acceptable standards are maintained for international consumption.

# 5
# The Need for Technological Advancement

## The Major Forms of Technological Change

THE acquisition and assimilation of technology by a manufacturing firm generally serve several major purposes, apart from initiating production for the market-place, either domestic or international. In the context of the domestic market, such acquisition will allow for the introduction of new products and services to satisfy increasing domestic needs. It also serves to enhance efficiency in the production of existing products through new processes, and the production of new products. However, all these activities require continuous technical adaptations and innovations within a firm; in the long run, it is these adaptations and innovations that will determine to a large degree the rate of economic progress.

When innovations are constantly diffused within a firm, they enhance product quality, reduce costs, and improve productivity, thus contributing to enhanced competitiveness, which is particularly important under an export-oriented regime. For a small, export-dependent economy such as Malaysia's, there are clear benefits to be gained in creating a conducive environment that will stimulate the nurturing of such a firm. When firms such as this expand in large numbers, there is a good chance that domestic industries as a whole will be at the leading edge of technology acquisition and adaptation. In theory, the gains will then be translated through improved productivity and ability to be competitive in the international market. However, in practice, the issues facing domestic industries in Malaysia are obviously more complex than this.

The level of technological capability of domestic industries at this stage does not compare favourably even with those of the Asian NICs and thus Malaysia ranks poorly among these economies on a number of measures of R & D performance. Part of the explanation lies with the relative lack of an industrial base, underdeveloped intersectoral and inter-industry linkages and economies of scale due to the small market size, as well as foreign dominance in major industrial sub-sectors. The relatively high tariff protection given to most of the import-substitution industries during the 1960s and 1970s tended to make them complacent within the confines of the domestic market; thus the lack of interest in being internationally competitive. An equally significant factor is the relatively slow pace of technology acquisition and adaptation by domestic industries, principally due to the lack of R & D personnel and facilities. Furthermore, the situation in Malaysia, as in most Third World countries, is one of technological duality in which highly modern, technologically sophisticated sectors, sub-sectors, or enterprises coexist with backward ones, in which low incomes, low productivity, and inefficiency are prevalent.

Being relatively resource-poor, the four Asian NICs, on the other hand, had in the 1970s and 1980s put substantial efforts into industrial technology upgrading through considerable investments in human resource development and relatively effective science and technology management policies. However, Malaysia, being well-endowed with its primary resources, has put its resources into the agricultural and natural resource sectors including rubber, palm oil, timber products, petroleum, and cocoa, which have hitherto contributed substantially to export and economic growth. Although such emphasis is a strength in itself, this could have detracted policy-makers from the complex task of laying a strong industrial and technological base, at least during the 1960s and 1970s.

Policy-makers have tended to concentrate on the primary sector given the strength of this sector in terms of its contribution to economic growth so that the more critical aspects of industrial development, such as the enhancement of indigenous technology, technology transfer, and the securing of backward and forward linkages, were not given the appropriate emphasis. At the same time, the role of the domestic private sector has not been very encouraging in this regard. For these reasons, domestic manufacturing firms are perceived as less competitive, thus leaving the MNCs as the focal point for enhancing the country's export-led industrialization. With this background, it is therefore critical that the rationale for technological enhancement be examined if Malaysia is to strengthen its industrial base so that domestic industries are as competitive as the MNCs in the international market.

At the outset it is important that the major forms of technological change be identified, and thereafter related to the Malaysian context. The following forms of technological change can occur within manufacturing firms (Fransman, 1986: 23–5).

### *The Search for New Products and Processes*

The literature on technology development has rightly emphasized that search activity, while costly, requires the firm to possess the necessary technological capabilities, which would obviously require highly skilled personnel and, perhaps, the support of a strong R & D facility. A certain amount of foundational knowledge is necessary for the firm to gain additional knowledge through its search for new processes and products.

*The Adaptation of Products and Processes*

Technology is to some extent implicit and location-specific. Accordingly, any transfer of technology will require a degree of adaptation. For instance, product development, including design and packaging, must be tailored, wherever appropriate, to local needs and preferences. However, in the literature a number of conditions have been singled out as being particularly important in the context of developing countries. These include the size and characteristics of domestic markets, the degree of competition in protected markets, the availability of different kinds of skilled labour, and the supply and quality of local inputs or resources.

*Improving Products and Processes*

Production within a firm must include the ability to adapt the production techniques and sale of products and processes to suit local conditions; this will inevitably involve their improvements in various ways. Such improvements may be incremental and minor, or they may be major improvements involving discrete jumps at a point in time. It is also important to note here that there is a cumulative significance of incremental improvements over time; such improvements, taking place within the developing countries, can be termed 'technological trajectories' and are unlike the technological improvements that are normally associated with manufacturing activities in the highly industrialized countries. In turn, these have resulted in technology exports in various forms from the NICs, such as South Korea and Taiwan.

*Developing New Products and Processes*

Here it is necessary to distinguish processes and products that are 'new' from the point of view of the firm, the industry, the country, and the world. Applying the earlier definitions at the level of the firm, the production of products or processes would be regarded as new to the firm if it involves technical change. In general, developing countries will very rarely produce products and processes that are in a fundamental sense new to the world. In contrast, in Japan, a new trend in technological innovation has emerged since the early 1970s, when increasing emphasis was put on the rapid expansion of R & D programmes and expenditures to produce new indigenous technologies. According to Hirono (1986: 135–50), the greater emphasis on R & D activities in Japan also reflected a depleting stock of advanced technologies in other industrial countries that were exportable to Japan, in part due to a persistent narrowing of the technological gap between Japan and these industrial countries.

*Basic Research*

While there is relatively little basic research carried out in developing countries, these activities do exist. This raises the question of how much basic research is justifiable and desirable at any time in any particular country. A further problem which has been discussed in the literature is the question of the relationship between scientific and technical institutions and production and exchange activities in the economy.

One of the critical issues in S & T policy formulation and implementation relates to the question of the depth of technical knowledge required in the country's various manufacturing activities at any particular stage of its industrialization. The importance of this issue becomes relevant when comparisons are made between the levels of technology needed and the costs of acquiring such technology. Also relevant is the question of appropriateness of any imported technology. Generally, developing countries do not need sophisticated levels of technological knowledge, particularly in the earlier stages of industrialization. As such, there is widespread agreement in the literature that technological activities in these countries are on the whole confined to three major forms: search, adaptation, and improving of products and processes. Thus, the development of domestic technological capabilities will also depend on the stages of economic development that developing countries are going through.

## The Technology Development Phases in Malaysia

Having identified the three major forms of technological change that take place in developing countries, it is useful to relate them to the following development phases of domestic technological capabilities. First, during the early industrialization phase, imported technologies from the industrial countries predominate and policy measures generally facilitate the process of technology acquisition within the industrial sector. This is where search activities for new products and processes become important, although generally it is the foreign investment component that plays a major role in this respect. In the second phase, when the industrial base becomes firmer, it involves the importation of technologies as in the preceding phase, but on a more selective basis depending on the needs of the industry sectors that are being promoted, while the strategies adopted would encourage a certain amount of indigenous technologies meant for the established industry sectors. At this stage, many of the production activities would also relate to product and process adaptation. In the third phase, the development of indigenous technological capability is generally adequate for selected industry sectors, and institutional linkage between the public sector and industry in terms of technology development emerges as an important mechanism to enhance domestic capabilities. The development of domestic capabilities at this stage would allow for product and process improvements.

During the first two phases, the processes of learning-by-doing and learning-by-adapting are important elements of technology development. This process of learning also means that the industrial work-force must be fairly well equipped to acquire technical skills, while shop-floor technicians, engineers, and technically trained managers are required in increasing numbers—to the extent that secondary and technical education becomes more important. For the third phase, learning-by-design becomes more significant so that design engineers play a more dominant role and advanced technical education is thus required. At the same time, S & T applications become more significant, especially in R & D activities in manufacturing firms, research institutes, and the universities.

At the general level, the domestic manufacturing sector has undergone the first and the second phases, and this is reflected in the increasing share of manufacturing in the country's output, while the transition towards the third phase is in its infancy. With increasing emphasis on technology development, it is hoped that Malaysia will move to the fourth phase during which the manufacturing sector will witness the maturing of indigenous technology. This will ultimately permit international exchange of technology at terms that are more favourable to Malaysia. However, there are varying degrees of achievement in the technological capabilities within the industry groups or sub-sectors, indicating the different levels of technology absorption and adaptation. In terms of the learning process, domestic industries exhibit varying degrees of learning-by-doing while learning-by-adapting has become more dominant, and in some sectors learning-by-design is becoming more feasible. The availability of an abundant supply of highly skilled technical manpower is an important precondition to reaching the maturity stage. Also needed is an effective institutional framework that will strengthen the capability to innovate.

The increasing technological competence of many of the domestic firms suggests that they are better positioned today to assimilate, and probably adapt, the technical and organizational components of imported technologies accompanying direct foreign investment than they were in the 1960s and 1970s. In this sense, they have to view any joint-venture with foreign capital or licensing arrangements with foreign technology suppliers as a means of complementing their own strengths.

Since the mid-1980s, domestic industries have been experiencing a technological phase in which they have substantially mastered the production or operating technologies necessary to run fairly sophisticated manufacturing enterprises ranging from food processing to electronics industries. Such production or operating technologies are generally associated with the initial steps in technology acquisition; that is, the ability to acquire operational skills to run a production facility efficiently. However, other more complex stages of technological advancement (including innovative technology, design, and engineering and consultancy technology) are yet to be fully developed locally. At the present stage of its industrial development—the early 1990s—the concept of technological advancement in Malaysia is often conceived as incremental, gradual, and achievable through many small modifications rather than being based on major breakthroughs as has been the case in most industrial countries. The existence of broad and popular education has probably been more significant to industrial expansion than the presence of a few scientific geniuses. Even so, such education must be able to adapt itself to the changing needs of industry.

The experiences of such NICs as Argentina, Brazil, India, South Korea, and Mexico since the mid-1970s have indicated the importance of expanding the domestic engineering and metalworking sector in furthering the development of technological expertise and industrial efficiency. These countries have lately begun to export capital equipment, turnkey projects, and engineering consultancy services. According to the World Bank, 'the competitiveness of these technology exports is founded on a history of learning, improving, and adapting technological process and products imported from industrialized nations' (World Bank, 1979: 66).

In the case of South Korea, Mukerjee (1986: 75) noted that:

> Korea's experience confirms that the growth of manufactured exports cannot be sustained over the long term without developing a capacity for technological innovation. But as the country moves into upmarket and technologically demanding products, it has to have independent research and development because countries from which technology is obtained may not be willing to part with it for fear of fostering competition. Less developed countries like Malaysia, which pose less of a threat to dominant industrial countries, may not face such problems immediately but they too may find that a competitive position cannot be established in the international market on the basis of licensed technology alone.

A higher phase of industrial development must also be complemented by the establishment of high-technology industries. But their development requires not only the provision of a highly skilled labour force but also a substantial increase in R & D allocations, particularly from the industry itself. It is in both these areas that manpower planning should focus its attention in attempting not only to develop the essential skills but, more importantly, in developing labour's capacity to innovate. It is the latter that will determine the economy's competitive edge *vis-à-vis* other countries (Foster, 1980: 136).

Basically, manpower planning up to the early 1990s has put little emphasis on the capacity to innovate. The lack of funds for purposes of R & D may be an important factor in this respect, but, more importantly, the existing industrial structure, arising from the role of foreign enterprises and the nature of existing investment incentives, encourages the adoption and importation of highly capital-intensive equipment from the industrial countries. This not only discourages the adoption of appropriate technologies in domestic industries but also negates the capacity to innovate as dependence on imported technologies is perpetuated.

Foreign MNCs in Malaysia seem to discourage the real transfer of technology and the development of domestic innovative capacity. There is a reluctance on their part to allocate funds for R & D activities in their local subsidiaries since most of these activities are controlled by their head offices in the industrial countries where extensive R & D facilities have already been established (Anuwar Ali and Muhd Anuar Adnan, 1990: 151–71; Ozawa, 1982: 7–53). The shortage of qualified and technically experienced scientists and engineers locally is an added impediment to the establishment of R & D facilities (Fisher, 1979: 88–103). Consequently, the top management of foreign MNCs tends to perform mostly management and organizational functions rather than innovative functions, unlike their counterparts in the industrial countries. In this respect, Nurul Islam (1988: 399–420) notes:

> Looked at from the point of view of the LDCs, international flow of capital and technology is an important index of their dependence on the developed world. The interdependence between the DCs and the LDCs is asymmetrical. While between the DCs such as the United States, the European Economic Community and Japan, there is a two-way flow of capital and technology, there are only one-way flows of capital and technology from the DCs to the LDCs. Since a predominant proportion of research and development expenditure in the advanced countries is undertaken by the multinational corporations, decisions for the generation and diffusion of the great bulk of world technology in the non-Communist world are centralised in their hands. Their operations do introduce considerable imperfections in the market and obstruct the competitive forces.

Such a situation has a stifling effect on the development of domestic entrepreneurship. However, in an industrializing economy, it is the domestic entrepreneurship that must play a more prominent role in combining the factors of production, developing new products, opening new markets and sources of supply, and introducing new production techniques (Sutcliffe, 1971: 109). The apparent lack of domestic entrepreneurship does not imply that there are insufficient individuals who can seize the commercial or industrial possibilities, but more often, implies that certain features of the economic environment inhibit entrepreneurial activity. For example, while technologies may be transferable from the industrial countries to Malaysia, the costs have often been prohibitive, and such costs will, to a large extent, be determined by the technological gap between Malaysia and the technology suppliers. The price paid for such technology transfer may also reflect the lack of technical, financial, and commercial expertise required for the acquisition of information on the technology itself and for the evaluation of the various technology options which may exist.

Many existing manufacturing enterprises do not provide a sufficiently relevant model to potential domestic entrepreneurs to stimulate the latter's independent activity. Large-scale foreign-owned enterprises, using techniques of organization and production methods of the industrial countries, cannot be regarded as good models as their techniques are so very different in character from the sort of industrial activity which characterizes the early stages of industrialization. In comparison to locally owned enterprises, foreign enterprises are generally larger, possess more experience in manufacturing activities, and have greater access to more readily available sources of capital and technical expertise. As such, they will have a competitive edge so that, in conjunction with various fiscal incentives and tariff protection provided by the host government, they can easily take up the most profitable opportunities in manufacturing to the detriment of domestic enterprises (Griffin, 1969: 125).

This creates an industrial structure which leaves to domestic entrepreneurship areas which contribute least towards an independent industrialization process, such as services industries, including trading, and small manufacturing activities. This tendency is reinforced by other structural features characteristic of open economies, such as Malaysia's, where a high demand for luxury consumer goods tends to ensure that the areas of economic activity of highest profit, into which domestic entrepreneurs are attracted, are found in commercial activities related to consumption goods.

Technology and R & D related activities in most developing countries, if there are any, tend to be almost exclusively of the incremental kind rather than of the frontier-moving or innovative type. To the extent that the latter occurs within these countries, it is generally the by-product of technology generated abroad rather than that based on domestic capabilities. However, it is important not to underestimate the cumulative significance of incremental technological change. This is a critical issue, although many observers in the area of technology development and change have often put greater weightage on major innovations. At the same time, it would be wrong to underestimate the significance of the forms of technological change occurring within the developing countries, although the task of identifying stages of progression is inherently difficult because of the possibility of making generalizations relating to the degrees of complexity in technological change. In this respect, Fransman (1986: 111–12) noted that:

> There is realisation that over time incremental technical change is usually of greater significance than radical frontier-shifting change. Accordingly, the story of technical change in the highly industrialised countries is by no means entirely one of breakthroughs by highly trained scientists and engineers. While such events certainly occur in these industrial countries rather than in Third World countries, far more typical, and more significant in the longer run, are incremental changes of the kind which appear in all countries.

Even in Japan, its

> technology achievements have mainly been founded on engineering design perfection—improvements in materials or components and quality control rather on radically new technological breakthroughs. In this, at the shopfloor level, technological improvements are outcomes of team work, accomplished in many cases by those who often form quality control circles—the shopfloor operatives and their foremen (Kang, 1986: 180).

This is also a reflection of the fact that operational decision-making in

Japanese firms is largely decentralized. Shop-floor workers are expected to solve local problems and, together with engineers, to produce innovative results. Many of the innovative ideas have actually originated from the shop floor (Ozaki, 1991: 60–1). Writing on the management of technological innovation based on the experiences of many US leading firms, Burgelman and Maidique (1988: 143) also observed that:

> Major systems innovations have been followed by countless minor product and systems improvements, and the latter account for more than half of the total ultimate economic gain due to their much greater number. While cost reduction seems to have been the major incentive for most of these innovations, major advances in performance have also resulted from such small engineering and production adjustments. Such incremental innovation typically results in an increasingly specialised system in which economies of scale and the development of mass markets are extremely important.

## The Nature of Technology Transfer in Malaysia

Although the term 'technology transfer' may give the impression of the existence of a donor and a recipient instead of a seller and a buyer, it may also be said that technology transfer simply means the act of purchasing technology from another country based on the mutuality of interests. Thus, in the case of the importing country, technology is not received free; it has a price. The importing country, for instance, needs to assess the appropriateness of the technology transferred within its socio-economic environment as well as the direct and indirect costs of such transfer by the technology suppliers, especially the MNCs (Kirkpatrick et al., 1984: 105–13). Given that the assessment capability itself is lacking, it is thus crucial that technology transfer to any developing country be considered in accordance with all the components of technology required to ensure that domestic industries are adequately prepared to assimilate imported technologies.

Accordingly, there are four components of technology transfer: techno-ware, human resources, information, and organizational structures. This may explain why mere importation of capital equipment, machinery, and plants from the industrial countries will not automatically lead to technology transfer in the developing countries. Effective utilization of these imports requires considerable investments in the development of the other components of technology (UN ESCAP, 1988: 44–5). However, it may also be argued that the nature of acquiring foreign technologies is such that technology transfer represents no more than an initial step in the exploitation of available knowledge. What is more critical is the ability to acquire what is termed 'technological mastery', that is, operational command over technological knowledge, manifested in the ability to use knowledge effectively and achieved by the application of technological effort (Dahlman and Westphal, 1981: 12–26). An important aspect of this technological mastery is the ability to adapt technologies so as to make them better suited to the domestic environment either by altering output characteristics to reflect local needs or by modifying input specifications to permit the utilization of locally available materials or resources.

An important aim in encouraging direct foreign investment is the expectation that it will facilitate the transfer of scientific and technological knowledge from the industrial countries. The acquisition of such knowledge is seen to be a crucial component in enhancing the rate of economic growth, especially within the manufacturing sector. The flow of technology transfer into Malaysia, as shown by the number of formal contractual agreements, has increased rapidly since the early 1980s. However, it must be noted at this juncture that these numbers are a simple indicator of technology transfer, and may not reflect the reality of technological acquisition within domestic industries. Indeed, it is difficult to measure with any certainty the complex nature of technology transfer. During the 1975–89 period, the Ministry of International Trade and Industry scrutinized and approved a total of 1,579 agreements. While the number of such agreements averaged less than 60 a year before 1980, it nearly doubled during the post-1980 period, reaching a peak of 198 agreements in 1989. This, to a certain extent, is an indication of the increasing effort towards industrialization as well as the substantial inflows of direct foreign investment into Malaysia, although the number of agreements approved in 1990 dropped to 156 (Malaysia, 1991b: 193).

Almost 50 per cent of all agreements approved since 1975 were in the form of technical assistance and know-how agreements. Another 20 per cent were in the form of management agreements and joint-ventures (Table 5.1). While this indicates that a high concentration of technology transfer was in the manufacturing sector, it also shows a shift away from the 'packaged' type of foreign capital and technology transfer process. This has been particularly apparent since 1985 when the number of trade marks and patents agreements increased substantially so that by the end of 1989 the share of such agreements was 14.0 per cent of the total. Japan accounted for the highest proportion (34.0 per cent) of all agreements signed during the 1975–89 period. The other countries that were relatively important were the United Kingdom and the United States, accounting for 13.1 per cent and 11.1 per cent of the total respectively (Table 5.2). The increase in the number of agreements signed with Japanese technology suppliers has been particularly marked since 1980, coinciding with the launching of the government's Look East Policy.

The importance of Japan as a technology supplier to domestic industries is further shown by figures on the Malaysian import of capital equipment or machinery, which constitutes an important element of technology transfer. Since the mid-1970s, around 35 to 40 per cent of Malaysia's machinery needs have originated from Japan. Only the United States, amongst the other industrial countries, appears to challenge the position of Japan in this respect. Even then, its share was

TABLE 5.1
Types of Agreement Approved, 1975–1989

| *Types of Agreements* | *1975* | *1976* | *1977* | *1978* | *1979* | *1980* | *1981* | *1982* | *1983* | *1984* | *1985* | *1986* | *1987* | *1988* | *1989* | *Total* | *%* |
|---|---|---|---|---|---|---|---|---|---|---|---|---|---|---|---|---|---|
| Technical assistance and know-how | 27 | 30 | 21 | 48 | 54 | 57 | 64 | 48 | 61 | 71 | 51 | 50 | 53 | 67 | 77 | 779 | 49.3 |
| Management | 12 | 7 | 7 | 11 | 13 | 13 | 6 | 10 | 13 | 10 | 6 | 10 | 5 | 7 | 12 | 142 | 9.0 |
| Joint-venture | 6 | 6 | 4 | 7 | 8 | 14 | 22 | 14 | 14 | 17 | 9 | 19 | 11 | 11 | 15 | 177 | 11.2 |
| Services | 12 | 5 | 1 | 12 | 3 | 6 | 7 | 2 | 7 | 2 | 1 | 1 | 1 | 2 | 12 | 74 | 4.7 |
| Trade marks and patents | 1 | 5 | – | 4 | 4 | 4 | 8 | 8 | 7 | 1 | 19 | 33 | 30 | 44 | 53 | 221 | 14.0 |
| Turnkey and engineering | – | – | – | – | – | 5 | 5 | 4 | 4 | 6 | – | 1 | – | 1 | 1 | 27 | 1.7 |
| Others[a] | – | – | – | – | 5 | 15 | 19 | 8 | 25 | 12 | 10 | 9 | 10 | 18 | 28 | 159 | 10.1 |
| Total | 58 | 53 | 33 | 82 | 87 | 114 | 131 | 94 | 131 | 119 | 96 | 123 | 110 | 150 | 198 | 1,579 | 100 |

*Source*: Ministry of International Trade and Industry, Malaysia.
[a]Others include supply and purchase, sales, marketing, and distribution.

TABLE 5.2
Agreements by Country of Origin, 1975–1989

| Countries | 1975 | 1976 | 1977 | 1978 | 1979 | 1980 | 1981 | 1982 | 1983 | 1984 | 1985 | 1986 | 1987 | 1988 | 1989 | Total | % |
|---|---|---|---|---|---|---|---|---|---|---|---|---|---|---|---|---|---|
| Japan | 22 | 21 | 7 | 32 | 21 | 32 | 35 | 33 | 46 | 39 | 33 | 38 | 37 | 54 | 87 | 537 | 34.0 |
| United Kingdom | 10 | 6 | 4 | 13 | 11 | 20 | 17 | 6 | 19 | 11 | 14 | 21 | 17 | 17 | 21 | 207 | 13.1 |
| United States | 6 | 4 | 1 | 9 | 8 | 11 | 14 | 10 | 18 | 12 | 13 | 12 | 12 | 22 | 23 | 175 | 11.1 |
| India | 3 | 5 | 8 | 7 | 5 | 5 | 4 | 4 | 4 | 2 | 6 | 1 | – | 2 | 2 | 58 | 3.7 |
| West Germany | – | 1 | 4 | 6 | 11 | 9 | 11 | 10 | 2 | 2 | 3 | 2 | 5 | 5 | 6 | 77 | 4.9 |
| Australia | 3 | 2 | 1 | – | 4 | 10 | 5 | 6 | 2 | 5 | 3 | 9 | 3 | 16 | 2 | 71 | 4.5 |
| Hong Kong | 1 | – | 3 | 3 | 2 | 9 | 2 | 3 | 2 | 7 | 4 | 7 | 7 | 5 | 9 | 64 | 4.1 |
| Singapore | 3 | 2 | 2 | 1 | 2 | 4 | 7 | 5 | 3 | 8 | 2 | 3 | 4 | 4 | 3 | 53 | 3.4 |
| France | 2 | 4 | – | – | 2 | – | 7 | – | 4 | 1 | – | 4 | 3 | 2 | 4 | 33 | 2.1 |
| Italy | 1 | 1 | – | 1 | 1 | 2 | – | – | – | – | 2 | 1 | 1 | – | 3 | 13 | 0.8 |
| Panama | – | – | 3 | – | 1 | 1 | – | 1 | – | – | – | 5 | 1 | – | – | 12 | 0.7 |
| Switzerland | – | – | – | 2 | 1 | – | 3 | 1 | 2 | 2 | 1 | 1 | 1 | 3 | 2 | 19 | 1.2 |
| Norway | – | – | – | – | 1 | 1 | – | 2 | 1 | 2 | 5 | – | 2 | 1 | – | 15 | 0.9 |
| South Korea | – | 1 | – | 1 | – | – | – | 2 | 4 | 6 | 1 | 3 | 1 | – | 3 | 22 | 1.4 |
| Others | 7 | 6 | – | 7 | 17 | 10 | 26 | 11 | 24 | 22 | 9 | 16 | 16 | 19 | 33 | 223 | 14.1 |
| Total | 58 | 53 | 33 | 82 | 87 | 114 | 131 | 94 | 131 | 119 | 96 | 123 | 110 | 150 | 198 | 1,579 | 100 |

*Source*: As for Table 5.1.

TABLE 5.3
Agreements by Industry Groups, 1975–1989

| | *1975* | *1976* | *1977* | *1978* | *1979* | *1980* | *1981* | *1982* | *1983* | *1984* | *1985* | *1986* | *1987* | *1988* | *1989* | *Total* | *%* |
|---|---|---|---|---|---|---|---|---|---|---|---|---|---|---|---|---|---|
| Electronic and electrical | 17 | 9 | 5 | 21 | 15 | 19 | 16 | 19 | 15 | 21 | 20 | 12 | 29 | 37 | 40 | 295 | 18.7 |
| Fabricated metal | 8 | 3 | 5 | 7 | 16 | 6 | 14 | 7 | 12 | 3 | 9 | 22 | 21 | 17 | 7 | 157 | 9.9 |
| Chemical | 3 | – | 4 | 19 | 8 | 11 | 21 | 5 | 15 | 17 | 16 | 15 | 18 | 29 | 27 | 208 | 13.2 |
| Transport equipment | 5 | 4 | – | 5 | 7 | 10 | 11 | 11 | 22 | 17 | 20 | 15 | 4 | 1 | 15 | 147 | 9.3 |
| Food | 4 | 7 | 2 | 2 | 8 | 14 | 15 | 4 | 21 | 6 | 10 | 8 | 8 | 16 | 21 | 146 | 9.2 |
| Textiles | 6 | 7 | 2 | 4 | – | 8 | 5 | 2 | 5 | 6 | 1 | 7 | 2 | 6 | 4 | 65 | 4.1 |
| Basic metal | – | 5 | 3 | 3 | 5 | 7 | 10 | 13 | 5 | 5 | 1 | 1 | 2 | – | 6 | 66 | 4.2 |
| Wood and wood products | 4 | 1 | 6 | 5 | 4 | – | – | 4 | 1 | 6 | – | 4 | 1 | – | – | 36 | 2.3 |
| Pulp, paper, printing, and publishing | – | – | – | – | – | – | – | – | – | – | 3 | 4 | 1 | 3 | – | 11 | 0.7 |
| Rubber and rubber products | 6 | – | 1 | 2 | 5 | 8 | 14 | 2 | 7 | 5 | 4 | 13 | 8 | 22 | 18 | 115 | 7.3 |
| Non-metallic mineral products | 1 | 6 | 1 | 1 | 7 | 5 | 4 | 16 | 9 | 17 | 7 | 7 | 12 | 4 | 10 | 107 | 6.8 |
| Hotel and tourist complex | – | 5 | 1 | – | 2 | 4 | 2 | 4 | 8 | 7 | 4 | 4 | 1 | 2 | 6 | 50 | 3.1 |
| Plastic | 1 | – | 2 | – | 3 | 5 | 6 | 1 | 2 | 7 | – | 4 | – | 2 | 6 | 39 | 2.5 |
| Others | 3 | 6 | 1 | 13 | 7 | 17 | 13 | 6 | 9 | 2 | 1 | 7 | 3 | 11 | 38 | 137 | 8.7 |
| Total | 58 | 53 | 33 | 82 | 87 | 114 | 131 | 94 | 131 | 119 | 96 | 123 | 110 | 150 | 198 | 1,579 | 100 |

*Source*: As for Table 5.1.

well below that of Japan. However, capital equipment or machinery alone does not constitute technology. It represents only that part of the technology embodied in hardware, while the remainder is comprised of disembodied knowledge which can only be acquired through indigenous technological effort, leading to technological mastery through human resource development. Most of the agreements approved during the 1975–89 period were in the electronics and electrical (18.7 per cent), chemical (13.2 per cent), and fabricated metal industries (9.9 per cent). These three industries accounted for more than 42 per cent of all agreements approved, thus reflecting the strong need for technology transfer in these areas (Table 5.3).

## The Existing Policy Framework and Approval Mechanism

The institutional and policy framework for regulating technology transfer in Malaysia has evolved over the years within the Ministry of International Trade and Industry. The institution of such regulations is not confined to Malaysia alone as other developing countries also regulate technology transfer. Thus, according to Marton (1986: 409–26):

> Despite the differences in form of regulation, such as by legal enactment, as in Mexico and Brazil, or by administrative guidelines, as in India and the Republic of Korea, the intent of regulation has been similar. In earlier phases, regulation aimed to reduce the costs of foreign technology and permissible maximum payments were established, apart from the assessment of the costs of equipment and maintenance inputs required to be imported as part of foreign technology usage. Subsequently, measurement of costs has also included various indirect costs arising out of restrictive provisions imposed by technology licensors.

Prior to 1968, manufacturing companies had only to submit technology transfer agreements to Bank Negara Malaysia (the Central Bank of Malaysia) under the Foreign Exchange Control Act in lieu of any royalty or technical fee remittances overseas. Since Bank Negara Malaysia was then mainly concerned with the amount of remittances, agreements were not subjected to a screening process. As indicated in the earlier chapters, the Investment Incentives Act of 1968 was basically designed to attract direct foreign investment by providing total or partial tax relief to companies involved in new manufacturing projects or in expanding into new products. Companies granted these incentives were subjected to a number of conditions, one of which required them to submit all agreements signed with foreign companies for approval.

Following the implementation of the Industrial Coordination Act of 1975, the Technology Transfer Unit (TTU) was established within the Ministry of International Trade and Industry for the specific purpose of screening all these agreements. Under the Act, all manufacturing firms requiring licence registration are also required to have any agreement signed with any foreign company approved by the TTU, whose main objectives are to ensure that: (i) the agreement will not be prejudicial to the national interest; (ii) the agreement will not impose unfair and unjustifiable restrictions on the Malaysian party; (iii) the payments of fees, wherever applicable, will be commensurate with the level of technology to be transferred and will not have adverse effects on Malaysia's balance of payments; and (iv) there will be meaningful transfer of technology.

A set of broad policy guidelines has been established by the TTU to monitor any agreement on technology transfer between a foreign licenser and a local licensee. The guidelines have, of course, been revised on various occasions to reflect the changing investment scenario and the changing needs of domestic industries. Prior to the establishment of the TTU, it had been claimed that domestic companies under licence generally had to bear a higher direct cost for technology as demanded by their technology licensers. The mode of calculating technical assistance and royalty fees seemed to favour the technology licensers. However, the TTU has insisted that all technical fees or royalty payments be based on net sales rather than gross sales, with the former being defined as gross sales less sales discount or returns, transport costs, insurance, duties, taxes, and any other charges.

Even if the direct costs of technology transfer were not excessive, domestic manufacturers might still be disadvantaged substantially if restrictive conditions were imposed by their foreign technology licensers. These might include restrictions on export outlets, the level of technology transferred, and domestic R & D activities. The TTU has, however, tried to minimize the negative impact of such restrictions so as to allow domestic manufacturers more flexibility to expand their operations and to become, in the long run, less dependent on their foreign technology suppliers. For instance, if the licensers insist on export restrictions, the TTU will demand that consent for sales outside the restricted territories should not be unreasonably withheld. At the very least, the TTU requires the technology licenser to allow its domestic licensee to export to other ASEAN countries.

A number of technology licensers insist on fixing the prices of the licensed products while others insist that the local licensees purchase all material inputs and components from them and at prices fixed by them. The second restriction appears to be more common. Such tie-in purchases strengthen the position of the technology suppliers, thus enabling them to maximize their gains by selling over-priced components, intermediate inputs, capital equipment, and spare parts. Furthermore, the obligation to purchase key inputs from the technology licenser enables the latter to monitor the activities of the local licensee, including a constant check on the production volume of the licensee, mainly for the purpose of determining technical fees.

In order to avoid such a situation, the TTU has laid down a number of guidelines regarding purchases of components or intermediate inputs. First, the domestic licensee should determine alternative sources of supply as far as possible. This implies the need for achieving a higher level of technological capability within domestic industries. Secondly, any clause binding the domestic licensee to purchase all imported components and supplies through the technology supplier should be avoided, unless no suitable alternative source is available. Thirdly, if such a provision cannot be avoided, the domestic licensee should stipulate that the prices are to be based on internationally competitive prices, with the manner of determining them described; that the most favoured licensee clause will apply to pricing; and that the prices charged will be the same as those paid by the technology licenser plus reasonable handling charges.

Even though the technology to be imported or developed can be identified and chosen, the shortage of highly skilled labour, including R & D personnel, may still prevent its efficient implementation. The scarcity of technically skilled labour has often been a constraint in the transfer of technology to the domestic industries. It is therefore imperative, from the government's point of view, that adequate training be provided at both the licenser's plant facilities and the local plant.

Despite the above stipulations, it is an impossible task for the TTU to ensure that there is real transfer of technology given the lack of experience and expertise within the TTU itself to assess the 'technology content' that is imparted to domestic licensees or local personnel, not to mention the arduous task of keeping track of the ever increasing number of companies which submit their technical assistance or joint-venture agreements to the TTU for approval. It is possible that the Ministry of International Trade and Industry as a whole was more concerned about the achievements of the New Economic Policy (NEP) objectives during the 1970s and 1980s, and was thus distracted from effective monitoring of the level of technology being transferred to domestic manufacturing enterprises. As emphasized by the Second Malaysia Plan (Malaysia, 1971: 1), the NEP aims

> at accelerating the process of restructuring Malaysian society to correct economic imbalance, so as to reduce and eventually eliminate the identification of race with economic function. This process involves the modernisation of rural life, a rapid and balanced growth of urban activities and the creation of a Malay commercial and industrial community in all categories and all levels of operation, so that Malays and other indigenous people will become full partners in all aspects of the economic life of the nation.

Such a situation is not only confined to Malaysia, but also applies to other developing countries where the monitoring of the terms of technology transfer or obligations by the technology suppliers has become an important issue. In India, for instance, although the government could exercise some pressure on the foreign technology suppliers to reduce the royalty payments, duration of the agreement, and lump-sum payments, once an agreement was approved there was a lack of monitoring on matters such as the production level of collaborated products, the licenser's adherence to the timing of the transfer of various components of the technology package, and the fulfilment of the export obligations. The Indian government thus did not seem to monitor the agreements and assess systematically the performance of the local companies acquiring the imported technologies. In view of this unpreparedness to ensure compliance with the conditions of the agreement, the companies also seemed to take their obligations lightly (Patvardhan, 1984: 43–102).

In this respect, the United Nations Centre on Transnational Corporations (UNCTC, 1988: 131) also noted that:

> Most developing countries have yet to develop sufficient expertise and skill in monitoring the performance of foreign investment and technology. This is perhaps due to the fact that monitoring is a complex and difficult process both on the conceptual and administrative planes. On the conceptual side, the difficulties arise in the determination and analysis of the kind of information to be obtained, especially with respect to the financial, technical and production aspects of the operations. On the administrative side, the difficulties arise with regard to the agencies responsible for monitoring their adequacy and skills. If a single agency is entrusted with the responsibility, it may find itself unable to cope with the volume of work, particularly with the growing number and variety of foreign enterprises. Monitoring may then be reduced to the mere gathering of statistical information. Such an agency may also lack the technical expertise necessary for a proper monitoring of all kinds of foreign investment and technology.

## The Policy on Remunerations for Technology Transfer

The cost of technology transfer to domestic industries is, to a large extent, influenced by the bargaining strength of the foreign technology licenser, the local company or subsidiary, or joint-venture partner, and, to a lesser extent, the Ministry of International Trade and Industry. Each of these parties has its own view of the value of the technology and its own preferences as to how it should be transferred. The Ministry, reflecting the interests of the local company, is desirous that technology be acquired at lowest cost, basing its calculations on long-term national interests rather than on the immediate returns to the firm. Such calculations include the social costs and benefits of each source of technology, the linkages with other industries, the use of domestic resources, the direct costs of technology (for example, royalties) as well as the hidden costs, such as possible overpricing by the technology licenser of inputs to the local licensee.

There are two basic modes of payment by which the transfer of technology is remunerated and agreed to by the government. The first is a lump-sum payment. This is only permitted in special cases where the technology can be fully and immediately transferred and absorbed by the licensee. It normally involves payments for less sophisticated technology or production techniques which do not necessitate any continuous flow of technical assistance from the licenser. Lump-sum payments are, however, disliked by technology importing countries because of the extra sum involved, especially at the initial or pre-operating stages, on top of the substantial investments needed for plant, machinery, and working capital. Furthermore, the payment has to be made before the licensee is really sure of the project being successful. However, after a case assessment by the Ministry, the latter may stipulate that the lump-sum payment be made in instalments.

The second method is the running royalty payments. These are linked to production or sales performance, and the computation is generally calculated as a percentage of 'net sales' or ex-factory price. This method is generally preferred by the Ministry, who usually fixes the rate at between 1 and 5 per cent of net sales. For less sophisticated technology and assembly operations, a royalty fee not exceeding 2 per cent of net sales is usually permitted. This is to ensure that such payments reflect the technology needs of domestic industries and their capacity to meet the costs demanded by the technology suppliers.

Royalties exceeding 2 per cent of net sales are allowed in the case of highly sophisticated technology (for example, a wafer fabrication project), projects involving high 'local content' (for example, if more than 50 per cent), export-oriented industries, heavy industries, and priority industries (for example, automotive component parts). Approvals since the mid-1980s indicate that the Ministry has in fact been fairly flexible in approving applications with royalty rates exceeding 2 per cent of net sales. Industries with the highest proportion of agreements approved above the 2 per cent level include food and beverages, tobacco, chemicals, rubber products, machinery, electrical machinery and appliances, and transport equipment. In the rubber products sub-sector, more than 80 per cent of the agreements submitted during the 1985–7 period were approved with more than a 2 per cent royalty rate, while in the electrical goods sector, 5 out of 44 cases were given royalty rates exceeding 5 per cent. In the rubber products sub-sector, more than 80 per cent of the agreements submitted during the 1985–7 period were approved with more than a 2 per cent royalty rate, while in the electrical goods sector, 5 out of 44 cases were given a royalty rate exceeding 5 per cent.

To a large degree, these observations indicate the extent to which the Ministry of International Trade and Industry has been flexible in its attempts to attract investments into the promoted industries and to accommodate the increasingly technological needs of these industries in the form of new machinery, equipment, and processes as well as management expertise and marketing know-how. An example is the resource-based rubber products sub-sector which has attracted substantial interest among foreign investors. With growing industrialization, this becomes inevitable because of the need for diversification and production of more technologically advanced products, where the availability of technology is relatively limited to a few MNCs. In such cases, technology owners have not always been willing to transfer their technology under the country's terms of technology regulation, particularly when this has been accompanied by controls over direct foreign investment.

Technology payments required for new and increasingly complex technologies may have to be significantly higher than the maximum rates prescribed in national guidelines for the manufacturing sector as a whole. The effective absorption of more complex and advanced technologies may also require a longer period than the duration prescribed in guidelines. The broader impact of the economic slow-down in 1985 and 1986 has also slowed the pace of technology inflow and has necessitated promotional measures and relaxations in regulatory guidelines. While regulatory controls continue to be applied for relatively mature and standardized technologies, the acquisition of the latest techniques in high-technology areas and sectors of special priority is increasingly being encouraged and promoted through various incentives, including exemptions from regulatory norms and guidelines.

In spite of the rapid increase in the number of formal contractual agreements with foreign firms, domestic manufacturers and their foreign counterparts must be made to adopt effective programmes to raise technological capabilities within their production facilities if the ambitious targets of the IMP are to be achieved. This has been noted in the IMP, albeit at a rather general level. Several broad policy recommendations have been put forward but there appears to be very little that deals specifically with the issue of increasing the effectiveness of technology transfer through direct foreign investment, especially via the MNCs.

Any future programme to produce a more effective mechanism for technology transfer must take into consideration some of the more important trends concerning this transfer. First, the bulk of the world's useful technology is owned by a relatively small number of private firms from the industrial countries. These countries confer proprietary rights on most forms of technology (for example, patents and trade marks), enabling them to impose obligations and conditions on the users of the technology. For instance, it is illustrative to observe that

> Science-based industrial corporations by virtue of their capital are able not only to 'buy' the best scientific brains and other resources for undertaking unlimited R & D, but also to purchase patents of patentees who do not have the resources to exploit their inventions. Through this process as well as consolidation, patent pools and 'the regulated patent production through systematic industrial research' corporations have been able to concentrate patents under their domain and thus expand their 'monopoly of monopolies'. In most cases they are able to dominate a given industry and this creates the very condition for its perpetual control. In such a case they control the 'main stream of inventive thought' in that industry and can clog any further developments without their consent (Yankey, 1987: 8).

Secondly, technology transfer may take the form of a package, with foreign technical expertise and capital equipment included. This makes it difficult to evaluate the price of each component of the technology and allows for costs, direct or otherwise, to be inflated. Thirdly, there is increasing competition among developing countries to attract direct foreign investments in general and high-technology investments in particular. Fourthly, developing countries, in view of their weak science and technology information systems, lack the technical, financial, legal, and commercial expertise required for obtaining information about the technology and for the evaluation of the various alternatives which may exist.

As in any other developing country, there is little evidence to show that technology transfer regulations in Malaysia have been used nationally to screen for more 'appropriate technologies', partly because technology transfer is processed by administrators lacking the necessary technology background (Segal, 1987: 1–32; in the case of India, see Patvardhan, 1984: 43–102). There is therefore a need for the upgrading of the country's technology assessment capability to ensure that the negative effects of technology transfer are minimized. Hopefully, this will be the case in Malaysia given that steps have been taken to strengthen the technical capability of the institutions involved in the processing, evaluation, selection, and enforcement of technology transfer agreements (Malaysia, 1991b: 193).

The modern concept of technology assessment includes not only the minimization of the above-mentioned effects, but also the maximization of positive effects of technology acquisition and the development of technologies consistent with the capability of domestic industries and the country's developmental objectives. In developing countries, where one can take advantage of the latecomer situation, technical assistance is simply a step towards the disciplining of technological progress to maximize the positive effects while minimizing the negative effects (losses or degradation) on the local environment. In the latecomer countries, lack of technological advancement is associated with the lowering of economic well-being, wastage of valuable resources, increased social disorder, and rising political instability. Thus, for developing countries, a proper strategy may be to emphasize the maximization aspects while not completely forgetting the minimization aspects.

While Malaysia's economic structure is being transformed with a larger industrial base, the major objective of technology assessment in the country must be perceived in the following context. First, an effective mechanism must be established to evaluate the appropriateness of technologies for transfer and adaptation. This involves the identification of existing technologies available in the industrial countries that are somewhat compatible and have scope for adaptation within the capability of domestic industries. Secondly, this must be followed by a proper selection of technologies for domestic advancement, involving the identification of those existing indigenous or foreign technologies that are consistent with national goals. Lastly, there must be control of inappropriate technologies for the protection of the environment. This involves the identification of corrective measures for both local and imported technologies. In all these aspects, both the public sector and private industry must play their respective roles.

An important consideration is therefore to institutionalize the use of technology assessment in development planning. The common practice in such planning is the selection of projects with implicit choice of technology. In the long run, it may be more beneficial if the technology selection process be made more explicit. This will involve a reorientation in the planning process requiring that technology assessment become a part of the social process of decision-making. Equally important is the micro-level decision-making process. In the development of its technological strategy, a manufacturing enterprise must make decisions in the areas of technology selection and embodiment, technology sources, competitive timing, level of R & D investments, organization and policies concerning R & D, and competence levels (Burgelman and Maidique, 1988: 234–5).

These observations suggest that, in spite of the increasing pace of technology development and its commercialization world-wide, the costs of technology transfer to the developing countries may be very high, particularly in those countries where the infrastructure for technology forecasting and assessment is weak or non-existent. This may also suggest that the scope for adopting usable technologies may be rather limited in practice, while in areas where technologies could be acquired by domestic industries, the licensers are in a position to exert monopoly rents either directly or indirectly through transfer pricing. Thus, the developing countries are in a weak bargaining position either because of the competitive forces described earlier or because of a lack of knowledge and information. This disadvantage is particularly felt by small-scale enterprises which are generally domestically owned and controlled. Even in the more developed countries such as Australia, the situation is equally complicated (Parry, 1988: 349–65). The high costs of technology transfer may detract the developing countries from acquiring such technologies at the desired rate. And when it is actually acquired, the technology might not be the most appropriate or applied with the maximum impact.

However, the South Korean industrialization experience indicates otherwise. With its domestic enterprises having considerable managerial and technical expertise during the period of export-led growth, South Korea has managed to reduce dependence on 'packaged' imported technology and instead has created a conducive environment for effective unbundling. Thus, from the mid-1970s technology licensing began to obscure direct foreign investment as a source of technology transfer, and licensing agreements were principally tied to joint-ventures, the vast majority involving wholly owned South Korean enterprises. The ability to effectively utilize licensed technical know-how therefore depends substantively on the absorptive capacity of the local licensees, as reflected by the existence of skilled human resources and R & D capability.

# 6
# State Intervention and Technological Change

## The Early Research and Development Initiatives

IT was only with the launching of the Fifth Malaysia Plan (1986–90) and the Industrial Master Plan (1986–95) in early 1986 that Malaysia started to pay greater attention to the development of industrial technology and to initiate policy guidelines for the national development of science and technology (S & T). There had been earlier efforts, but most of them had been *ad hoc* in nature, and thus there was little focus and sense of direction. General research activities, however, date back as early as 1879 when the Forest Research Institute was established, followed by the Institute of Medical Research in 1901, and the Rubber Research Institute (RRI) in 1925. The RRI was set up with the aim of conducting research into all aspects of natural rubber cultivation and latex production, the development of new forms and uses of rubber, as well as technological research in the processing and manufacturing of natural rubber.

During the post-Independence period, consistent with the country's drive towards economic diversification, research efforts became more intensive, particularly with respect to agro-based activities. The Malaysian Agricultural Research and Development Institute (MARDI), established in 1969, was designated with these tasks, while the Palm Oil Research Institute of Malaysia (PORIM) was established in 1979 to spearhead research activities on palm oil. While MARDI's main objectives are to intensify research activities and programmes to assist farmers to increase farm productivity, to develop new technologies in farming, and to intensify research in food processing and preservation to maximize the utilization of agricultural products, PORIM's objectives are related to research on palm oil aimed at improving the efficiency of extraction and refining and end-use research to increase the proportion and improve the performance of palm oil in existing uses, such as edible and inedible products, or in new uses.

In conjunction with the Rubber Research Institute, the agricultural emphasis was on searching for new clones and high-yielding seed varieties of *padi*, rubber, and oil palm. It was also during this period that the importance of S & T was officially acknowledged although most of the efforts were directed towards agro-based activities. Perhaps the most significant development has been the consolidation of S & T related policy formulation and implementation through the establishment of the Ministry of Science, Technology and the Environment (MOSTE) in 1976, which had its origins in 1973 with the creation of the Ministry of Technology, Research and Local Government. The Ministry's role is complemented by the National Council for Scientific Research and Development (NCSRD) (established in 1975), whose primary function is to advise the government on scientific and technological matters.

In 1975, the merging of the Standards Institute of Malaysia and the National Institute of Scientific and Industrial Research to become the Standards and Industrial Research Institute of Malaysia (SIRIM) marked another significant development in terms of industrial R & D and technology development. Placed under the jurisdiction of the MOSTE, its functions are mainly related to the promotion and undertaking of industrial research and to assisting industries towards industrial efficiency and development. Apart from SIRIM, the Coordinating Council for Industrial Technology Transfer was formed under the MOSTE in 1977 to co-ordinate the activities of various agencies dealing with technology transfer so as to accelerate the industrialization process.

## Managing Technological Change

A critical problem that governments of developing countries such as Malaysia face in managing technology development is fostering technological effort that leads to the most efficient use of the available domestic resources. The rapid economic and technological changes at the international level and the widening technological gap between the industrial and the developing countries make it imperative that the state intervenes to effect the technological change needed for a sustainable industrial growth. Managing this change has become all the more critical given the increasing complexity of S & T as well as the rapid pace of innovations emanating principally from the MNCs of the industrial countries. At the same time, this constraint becomes less manageable since the technological input from domestic industries is rather weak.

The clusters of technologies and innovations in the industrial countries inevitably have a profound impact on manufacturing processes world-wide, and this has altered international trade relationships in various ways. Firms or industries which are capable of initiating innovations benefit through the marketing of new and improved products and processes and through the use of improved manufacturing techniques which the technologies provide. In the industrial countries, and to a lesser extent within the NICs, these developments offer possibilities for creating new industries, modernizing old ones, and generally restructuring the industrial base, thus leading to a competitive standing for these countries in world trade.

In these countries, it is evident that technology development is usually generated by the private sector, while the state is more concerned with projects which involve large capital outlays and long gestation periods or which possess numerous economic externalities. For example, most of the basic research in the areas of defence, atomic power, outer space, oceans, the environment, and national health come under the purview of the state. In fact, since the 1980s most industrial R & D and innovation and government policy in the most advanced countries have been focused on three clusters of technologies: information, materials, and biotechnology (OECD, 1985: 56–8).

The most advanced and widely diffused of these clusters is information technology which appears to be in the midst of a cumulative and virtually self-generating process of innovation. The basis of this technology is mainly micro-electronics and its data-processing capabilities. Materials technology is also advancing across a wide front including metals, alloys, composites, ceramics, and polymers. R & D is aimed at developing materials with specific properties suited to given conditions and requirements, through combinations of new and improved materials, design, and processing techniques. Biotechnology is now moving from the laboratory into industrial development and application. The number and variety of applications can be very large and diverse, including health care, food and agriculture, energy, and pollution abatement.

It is, of course, impossible to predict the directions these technologies will take, the combinations they may form, the pace at which they will develop and diffuse, and the impacts they may ultimately have on both the industrial and developing countries. Nevertheless, these technology clusters have several features in common which are critical for S & T policy in the future. First, the technologies are intertwined with and dependent upon science, particularly current research of a multi-disciplinary nature. The close coupling of science and technology, the rapid translation of research into application, and the merging of scientific disciplines into generators of new technologies enlarge the scope for innovation and quicken its pace.

Secondly, each technology is at an early stage of development and diffusion. Although the technologies are emerging and evolving rapidly, their development and impact lie mainly in the future. Thirdly, the development and exploitation of the technologies will require enormous investments, particularly in R & D infrastructure and human resource development. Investments on a large scale will be needed for continuing fundamental research to nurture the technologies and provide technical alternatives. This will require the education of scientists and engineers, often in curricula which have yet to be formulated, and the establishment or strengthening of institutions to support it.

In contrast, in the developing countries where the technology base is still embryonic and industries still lack the capability to develop industrial technology, the state must play a leading role in all these areas. The failure of market mechanisms to activate any concerted effort towards this kind of change makes it all the more important that the state be involved. Furthermore, the output of R & D activities has some of the attributes of a pure public good. Undertaking research is an extremely risky enterprise, and currently the community has not developed adequate risk markets to shoulder part of the inventor's risk. This also means that the government's role becomes crucial given that technology development requires substantial infrastructural support of all kinds, including education and training, technical extension services, development of public–private sector linkages, and a legal framework (that is, patent laws) to enforce property rights and maintain secrecy. Without the active encouragement and financial support of the state, the amount of R & D activity in society is likely to be less than optimal.

For developing countries such as Malaysia, where direct foreign investment has been a major player in manufacturing activities, it is even more difficult to manage this technological change in view of the rapid pace of technological changes and innovations which require shorter response time, from product development to marketing. Moreover, there is increasing specialization and complexity of information and management systems, bringing with them the need for proper interfacing of manpower and disciplines and integrating critical skills from various sectors of the economy. Furthermore, the new emerging technologies are rapidly becoming pervasive, and these will not only redefine traditional markets and spawn new ones, but will also affect working and living conditions of society at large.

While it is crucial that the state provides the leadership to enhance technological change, it is equally important that the state and industry collaborate closely. Up to the early 1990s, however, such collaboration in Malaysia has not been as extensive as in many industrial countries. The experiences of Japan and, to a lesser extent, South Korea, bear testimony to the necessity for such a collaboration. In the case of South Korea,

> the government has itself been engaged in providing research funds in areas of national interest. The National Project for R & D, started in 1982, aims to promote R & D in high-technology fields and stimulate the development of appropriate technology. In the project, two types of arrangements are possible. First, projects directly related to public interest, such as energy and natural resources, are funded entirely by the government. These government-funded projects are carried out by public research institutes. Second, projects with some commercial value are partially funded by the government and are called joint projects. Joint projects are usually initiated by private firms with the government providing a subsidy, which varies in extent depending upon the nature and type of research (Yoo, 1989: 80–104).

But more significantly, in both Japan and South Korea state intervention in the economy has been more extensive and readily visible than in most other industrial economies and has occurred at both macro and micro levels. The Economic Planning Agency of Japan and the Economic Planning Board of South Korea have both initiated economic plans which were somewhere between indicative and command planning, stipulating priorities of both governments and indicating to industry the direction in which the governments want the economy to move (Blumenthal and Lee, 1985: 221–35; Shahid Alam, 1989: 233–58). However, it must also be emphasized that such capabilities have been nurtured through the strengthening of the planning machinery within both countries, without which it would be almost impossible to indicate priorities or directions.

In Japan, an increasing emphasis on the development of new indigenous technologies has, since the early 1970s, intensified the need for greater government involvement in promoting R & D in private industry and in co-ordinating efforts to accelerate the pace of technological innovation. In addition to providing financial assistance and tax incentives to promote domestic technological innovation, the Japanese government has been placed under enormous pressure to expand its own R & D activity by way of increasing government R & D expenditures, expanding existing government and semi-government R & D laboratories, universities, and agencies, and setting up new organizations in strategic fields (Hirono, 1986: 135–50). At the same time, the Japanese government and private industry have initiated complementary programmes for the training and development of industrial manpower under broad guidelines set by the government. Although the latter does assist in technical training programmes, most vocational training is done by industry.

At the industry level, on-the-job training programmes are found to be a source of motivation for employees to achieve the companies' targets or goals, apart from providing basic knowledge of the companies (Asher and Inoue, 1985: 23–6). Even in Hong Kong, where state intervention is kept to the minimum, government policy since the late 1980s has shifted from a posture of non-involvement to one of increased intervention, as the promotion of high-technology and skill-intensive industries is considered crucial for enhancing technology transfer, innovations, and productivity improvement (Ng, 1987: 467–78). In the United States, state governments have been urged to take an active part in promoting university research to fulfil the need for innovation; this is done with the formulation of long-range strategies for the utilization of S & T resources, the identification of needs and opportunities for innovation, and the fostering of greater university–industry research interaction (Lindsey, 1985: 85–90). All these examples exemplify the importance of the state not only in providing the lead that industry needs but also in ensuring that both the public sector and industry interact closely to provide the momentum for R & D activities.

In the Malaysian context, where state–industry collaboration is still in its infancy, S & T implementing agencies need to be cautious about their degree of involvement in the day-to-day implementation of policy measures. While the state must be the prime mover in establishing technological priorities, its involvement, especially with regard to decision-making at the firm level, might be seen as unwarranted interference. However, one must also note that technological *laissez-faire* in the 1980s was as much a thing of the past as was economic *laissez-faire*. In most countries these changes in approach are reflected both at the political and administrative levels through the establishment of an institutional machinery to assess the impact of technology and to create awareness for greater public participation in decision-making with regard to S & T.

The state, therefore, plays a role in a rather broad way and has the task of both advancing and directing S & T. Of the innumerable approaches in which state agencies may affect technical innovation, some of the most important are through the creation of demand, subsidies to firms, regulation, technical and scientific infrastructure, support for small- and medium-scale manufacturing firms, and a banking system well adapted to the needs of innovative industries. The most important role of the state is thus to establish an environment that stimulates firms and specialized technological agents such as engineering firms, intermediate goods producers, and capital goods suppliers to engage in ongoing technological efforts and to upgrade technological capabilities that will ultimately improve productivity and overall economic performance. The state can also intervene directly to induce choices of techniques that are socially most appropriate, foster imports of technology on the best possible terms, and stimulate the development of new technology-based firms (Dahlman and Ross-Larson, 1987: 759–75).

The experiences of the Asian NICs with different policy regimes and degrees of state intervention indicate that the best approach lies somewhere between limited state intervention, through simply providing a broad policy environment, and a proactive state intervention through direct R & D support. On the one hand, extensive state intervention in the production system tends to introduce rigidities incompatible with the requirements of the new competitive environment, unless state enterprises have the autonomy in strategic decision-making. At the same time, public R & D institutes, as well as the universities, run the risk of being insulated from the needs of industry and thus may produce results which have limited commercial potential. On the other hand, the minimum state intervention formula ignores certain potential external economies to which new technologies give rise.

The pace of development and international competitiveness of industries in the developing countries, and particularly in the NICs, is determined by their technological competence and capabilities to identify, select, adapt, diffuse, and improve modern technologies to suit domestic conditions as well as the changing international environment. The NICs have so far achieved rapid economic growth largely because of such technological competence. For them, continuing efforts to acquire and strengthen this competence have become an integral component of their industrial planning process. Hence the significance of appropriate state policies relating to S & T development.

Given that Malaysia's low technological base is impeding efforts at industrial broadening and deepening, there is now a realization that it is essential to close its growing economic divergence from the industrial countries (and also from the Asian NICs) by narrowing the technology gap. The dilemma is that the principal sources of new and improved technologies are in the more industrialized countries. The fact remains, too, that the technology development of any country wishing to industrialize invariably proceeds with imported industrial country technologies, and progresses through the development of domestic variants of these imported technologies to the final goal of technological self-reliance.

However, little can be expected from imported technologies in the absence of a domestic capability to modify and improve them for local applications. Even then, this must be done within the locally owned enterprises. Therefore, to achieve viable results from the technology acquisition process, substantive efforts must be made to assimilate and adapt imported technologies. These efforts will, of course, depend on human resource development, that is, the upgrading of engineering and technical skills through the formal education system, especially in the long run. However, in the short and medium term, skill enhancement at the firm level through in-house training and apprenticeship schemes ought to be rapidly increased.

The experiences of the Asian NICs indicate that they have been especially successful in utilizing international trade as a vehicle, not only to accelerate economic growth but also to enhance technological capability. High volumes of production for export have permitted them to reap substantial benefits through economies of scale and the learning process. The sort of learning involved, however, is localized in a limited number of activities associated with low cost manufacturing of standardized products. Continued dependence on such products to sustain export expansion is becoming increasingly difficult, and may erode their competitive edge. To be able to move to a different level of competitive advantage based on product or quality enhancement will therefore require a new set of skills which demand additional investments in imported technologies as well as in complementary training and R & D activities. At the same time, considerable investments in marketing and distribution will have to be made, and these could be more substantial than investments in R & D given the global competitive environment. Even then, there is no guarantee of success as the international environment is far less favourable at present for continued strong export expansion than it was in the 1960s and 1970s when the Asian NICs first embarked on their export-led growth.

## The Private Sector: Lacking Technological Input

The state as an initiator of industrial development will have to take the lead in promoting R & D by industry through the creation of an appropriate environment for such activities by providing fiscal incentives, establishing an effective delivery/information system, and creating special R & D funds. Market forces alone may not be sufficient to generate technological advancement and industrial growth in the 1990s, particularly given the export-orientation of the manufacturing sector. This has been recognized in countries such as Japan and South Korea where the state interventionist policies have greatly shaped the extent and depth of their respective technology development. However, once a high level of technology development has been achieved as in the industrial countries, research institutes are no longer the only source of new emerging technologies about production processes and products. In fact, industry itself can play a leading role in fundamental research as well as in developing applications of such technologies. There is no doubt that resources required for effective R & D activities are often substantial and beyond the reach of individual research institutes or universities. Furthermore, the gestation from basic research to commercialization has become shorter, thus giving commercial justification for industry involvement. The consequence is that advances in S & T are increasingly driven by industrial requirements and are rapidly incorporated into business strategies.

In Malaysia, however, the private sector does not have as yet a tradition of industrial technology development, although there are some exceptions. According to the Sixth Malaysia Plan (1991–5), there is extensive reliance by the private sector on foreign sources of technology, while in-house R & D activities are not significant. Domestic adaptations and modifications of imported technologies in the form of new and improved products and processes are not substantial either (Malaysia, 1991b: 192). This situation arises because both the earlier import-substitution and export-led industrialization policies were dependent on direct foreign investment, and thus technologies and technical expertise were imported. Since it responds to market signals and its planning horizon is relatively short, the private sector, in fulfilling its corporate interests, does not have the incentive to undertake projects in industrial technology development whose returns may not be immediate. The market, therefore, fails to work as an adequate mechanism for allocating resources to technology development or R & D activities because decision-making at the firm level is based principally on the profit motive.

An increase in R & D expenditure commensurate with domestic absorptive capacity needs to be given priority in any future technology development strategies. Apart from the state's policy direction and active support for R & D activities, it is vital that a significant portion of this increase be undertaken by the private sector. This is in line with the Fifth Malaysia Plan's objective to increase the private sector's role in achieving higher value-added from the exploitation of the country's resources. In the past, private sector involvement in national R & D has been minimal (Malaysia, 1986a: 269). It is thus instructive to note the remarks made by Yoshihara on the role of domestic industrialists *vis-à-vis* their counterparts in Japan:

> Unlike South-East Asian capitalists, Japanese capitalists did not depend on foreign companies for the layout of their factories or installation of machines. At that time, there was no concept of 'plant-export', but now a South-East Asian capitalist can essentially buy a plant from a foreign company as a turnkey project. If his staff do not know how to operate it, he can enlist a foreign engineering company to give the necessary training before the plant is completed; and if this is not enough, he can hire its technicians to stay on at the plant. Even if this is unnecessary, he may later have to depend on the engineering company or the supplier of the machines when something goes wrong, for his technical staff may lack the ability to make complicated repairs (Yoshihara, 1988: 112–13).

However, in the case of South Korea and Taiwan, whose major sources of growth are already oriented towards the knowledge-based industries, the private sector has become the spearhead for such a development. While these two newly industrialized economies employ foreign technologies to their advantage, especially in managing supplier contracts and technology agreements or strategic alliances with foreign multinationals, they are also moving towards innovative R & D and developing software programmes and interactive networks that operate in Chinese and Korean language characters. Since neither of them has sufficient R & D resources to accomplish high technology on their own, many of their domestic corporations are stepping up formal linkages or strategic alliances with foreign high-tech firms (Crawford, 1987a: 10–16; for Taiwan, see also Liang and Liang, 1988: S67–S101).

The experience of South Korea in this respect during the 1986–90 period has been most illustrative. In the case of the electronics industry, for instance, South Korea has become the world's sixth largest in terms of output after the United States, Japan, Germany, France, and Britain. Across a wide spectrum of electronic products, South Korea now has a significant presence, which to a large extent has been spearheaded by its corporate sector, especially the four largest conglomerates, Samsung, Hyundai, Lucky-Goldstar, and Daewoo (Clifford, 1991: 66–7).

Similar trends are being observed in Singapore where private industry has become increasingly conspicuous in R & D activities. In 1988, for instance, the main driving force in these activities was private industry, accounting for 60 per cent of R & D expenditure and 44 per cent of R & D manpower (S. K. Lee, 1989). There is no doubt that advancement in high-tech industries (including optics, biotechnology, electronics, new materials, and information processing) will require substantial cross-fertilization of ideas among firms and even among different industries through the development of networking on technical information and personnel. This, as manifested in many of the industrial countries, is bound to facilitate cross-germination of ideas and inter-industry R & D (Ozaki, 1991: 61). It is in this sense that industry will have to play a bigger role in the development of industrial technology in Malaysia.

## An Affirmative Role of the State

It is appropriate that the state, especially in countries seeking to industrialize themselves, seeks to strengthen its grasp on S & T development in terms of policy direction and direct involvement. It has often been emphasized that technology transfer into the domestic manufacturing sector can be effected by careful planning and intervention rather than as a response to invisible market forces (Tuma, 1987: 403–7). There is a need for a well-defined industrial strategy integrating technology development with economic planning. This is an important precondition to the implementation of any action plan for technology development, and one which requires an affirmative role of the state.

In fact, as stated in the Second Outline Perspective Plan (1991–2000), one of the major development thrusts during the period will be 'making science and technology an integral component of socio-economic planning and development, which entails building competence in strategic and knowledge-based technologies, and promoting a science and technology culture in the process of building a modern industrial economy' (Malaysia, 1991a: 5). In this respect, agencies such as the Economic Planning Unit, the MOSTE, and other related ministries must play an important role as they are responsible for overall economic and S & T manpower planning.

Policy direction and the state's direct involvement can be perceived within the context of two major elements, that is, to increase political commitment and public awareness with respect to S & T development and to minimize the widening technological gap between Malaysia and the industrial countries. In this respect, the state should strive to achieve a number of important objectives. First, it is imperative that there is a commitment to S & T development at the highest political level so that an awareness is created on the importance of S & T for the future expansion of the economy. Such a political commitment would allow every level of society to appreciate and understand the need for S & T.

Secondly, in view of the need to stay competitive world-wide, the country should have the capacity to manage technological change. This obviously requires an appropriate environment that will foster technological self-reliance in selected or priority areas. As a precondition, there must be a high degree of indigenous technological capability. Thirdly, planners should enhance the process of integration between strategic S & T planning and national economic planning. This requires the strengthening of organizational structures and linkages with regard to policy development, allocation of funds and other resources, and effective R & D delivery mechanisms. Fourthly, there is always a need for a forum for a continuing dialogue between all interested parties in the development of S & T which include channels for linkages between the universities, research institutes, and industry.

During the Fifth Malaysia Plan period (1986–90), the state provided general guidelines on the proportion of resource allocation for basic,

TABLE 6.1
Distribution of Research Allocation to Types of Research by Different Categories of Research Institutions (percentage)

| *Research Institutions* | *Type of Research* | | | |
|---|---|---|---|---|
| | *Basic* | *Applied* | *Development* | *Distribution* |
| Universities | 40 | 50 | 10 | 22 |
| Government research institutes | 10 | 35 | 55 | 52 |
| Private research bodies | 5 | 20 | 75 | 26 |
| Total allocation | 18 | 35 | 47 | 100 |

*Source*: Malaysia (1986a).

applied, and developmental research, giving an overall ratio of 18 : 35 : 47 respectively (Table 6.1). The universities were supposed to play a major role in basic research, while the public sector research institutes were to concentrate on applied and developmental research. The private sector, on the other hand, was encouraged to utilize and commercialize the research results through its role in developmental research. This was in line with the recognition that indigenous technological capabilities will become increasingly essential for sustaining the industrialization process in the future. The respective roles of the state, research institutions, and industry have also been aptly emphasized by the Fifth Malaysia Plan (1986–90) (which for the first time devoted a full chapter to science and technology):

> ... the current uncertainties in the international environment, coupled with the difficulties in maintaining the competitiveness of manufactured products, necessitates an even greater role of S & T in the industrialization programme of the country. A strong base in S & T and vigorous support of R & D will, therefore, be crucial. During the Fifth Malaysia Plan period, the role of S & T as an effective tool of development will be further intensified, especially in the light of emphasis on increasing agricultural productivity and intensifying resource-based industrial development as well as expanding the manufacturing base to include heavy and high-technology industries. Greater private sector involvement will be encouraged. Opportunities will be provided for interactions between both primary and secondary industries and R & D institutions, including universities. Further efforts will be directed to increasing centralized planning and coordination in research programmes, strengthening the existing infrastructure for S & T management, expanding education and training in S & T, improving the technology transfer mechanism as well as encouraging and facilitating the utilization and commercialization of research results (Malaysia, 1986a: 261–2).

To facilitate all the above changes, it is crucial to adjust and improve upon the existing S & T infrastructure within a relevant time frame. It is useful to distinguish several key elements to the S & T infrastructure, such as: institutions engaged in basic scientific research, including research institutes and the universities; industrial research and R & D laboratories; engineering-based and new technology-based firms; mechanisms for financing innovative endeavours; and institutions for education and training of skilled personnel.

To appreciate fully the strengths and weaknesses of the country's S & T infrastructure, it is not enough to consider each of these components in isolation but also the dynamic interactions among them. In particular, the linkage or interfacing between basic research and industrial R & D would seem to be critical. It is also necessary to strengthen indigenous S & T infrastructure in terms of its collaboration with the international S & T community.

As indicated in the preceding sections, policy-making and implementation require a well-defined set of strategies for industrial technology development and for establishing clear guidelines to implement them at all levels of the national administrative machinery. The establishment of an industrial technology network at these levels and between the public sector and industry as well as the universities is most desirable in order to create an effective arrangement to identify common needs, and to locate and adapt appropriate technologies. Networking provides the institutional framework within which innovations can occur, stimulates the aggregation of common needs, and promotes collaborative problem solving. More importantly, these functions must be complemented by an effective human resource development programme to achieve the level of industrial expansion needed by the country. Such a programme is desirable because of the need to enhance creativity through the education system to foster greater manipulative skills and technical aptitude.

## Developing a National S & T Framework: The Role of the State in the 1990s

Since the national framework for S & T development was only instituted in 1986 with the launching of the Fifth Malaysia Plan (1986–90), it still lacks the effectiveness that is essential for enhancing domestic technological capabilities. As such, the growth of the industrial sector has generally occurred within an environment that is not complemented by a concomitant expansion in indigenous technological capability through the supply of highly skilled manpower. Consequently, the growth in manufacturing output has been accompanied by increasing dependence on technology imports. At the same time, the linkages between the public sector and industry are rather weak. This does not mean that there is a total absence of a S & T management system, but that the system is diffused because of the decentralization concept of management adopted.

Agencies such as science-related agencies, legal agencies, and other institutional arrangements in S & T tend to operate within their defined domains. Except in a few cases, there is little interaction between the research institutes and the higher institutes of learning because the mechanisms for such linkages are not properly thought out. Likewise, the R & D agencies operate without the benefit of an overall national development perspective. Since the present system of S & T management is substantially decentralized in approach, the Fifth Malaysia Plan (1986–90) proposed that a

> more centralised planning and co-ordinated implementation approach will be pursued with a view to achieving higher productivity in R & D. Towards this end, a comprehensive review of the science policy organisation as well as the legal and institutional arrangements in S & T will be undertaken. The NSCRD, as the major science policy organisation, will be strengthened to provide effective intersectoral jurisdiction in planning and management, while an independent mechanism will be provided for evaluation and assessment (Malaysia, 1986a: 269).

As indicated earlier, at the apex of the S & T framework is the MOSTE, which is assisted by the NCSRD. The principal objectives of the MOSTE are to develop and promote the expansion of S & T with the aim of improving the quality of life; formulate research; plan and ensure that the extensive application of S & T does not give rise to adverse effects; ensure that material development through S & T does not pollute the environment and destroy wildlife and plants; and integrate physical development through S & T with human and individual development in order to reduce undesirable conflicts and stresses that may arise from environmental and technological changes. If all these objectives are to be achieved, the MOSTE must certainly be fully equipped with enough manpower and expertise to plan and implement related strategies as well as have sufficient expenditure allocation.

While the general objectives of the NCSRD are to ensure that scientific research activities are geared to national development needs and goals, its main functions include formulating a Science Policy for the nation and undertaking an innovative role in relation to science for the progress and modernization of society; serving as the national scientific consultative and advisory body to the government; identifying R & D activities consonant with the nation's development objectives; initiating and co-ordinating R & D activities of the nation and ensuring maximum utilization of resources; developing the country's manpower potential for R & D activities; and promoting a free interplay in R & D between the public and private sectors.

While the role of the MOSTE is principally to promote S & T development and its infrastructure to enhance the contribution of both the industrial and non-industrial sectors, the NCSRD's role is basically advisory and its focus is on R & D activities. The direction of S & T development can be examined within the context of a three-tier institutional structure:

1. Cabinet Level: The highest level of S & T decision-making and control is the Cabinet. To support the Cabinet, the NCSRD was reconstituted and a Cabinet Committee on Science and Technology was established in 1990.

2. Advisory, Planning, and Coordination Level: Concrete policy proposals are initiated by the MOSTE and the Prime Minister's Department. The MOSTE is advised by the NCSRD while the Prime Minister's Department has a Science Adviser. The activities of other ministries such as the Ministry of International Trade and Industry, Ministry of Finance, Ministry of Education, and the Ministry of Labour and Manpower complement the institutional structure for S & T policy and development.
3. Implementation Level: The implementation of S & T and R & D policies is the responsibility of three groups of institutions—public sector institutes, the universities, and the private sector.

A number of concrete measures have already been taken to enhance industrial R & D, and these include the following:

1. The establishment of the Malaysian Institute of Microelectronic Systems (MIMOS) in 1985 to initiate R & D in software design and computer application.
2. The establishment of the Coordinating Council for Technology Transfer at the end of 1986 as part of the institutional support for industries to study, analyse, and monitor programmes on technology and R & D.
3. The establishment of a Technology Transfer Centre in SIRIM to provide information, extension services assistance in acquiring foreign technology patent evaluation, and training courses.
4. New plans to establish a national laboratory accreditation system where quality assurance and testing, physical evaluation, and safety tests would be conducted.
5. Plans for a Centre for Industrial Excellence.
6. Plans for a National Centre for computer assisted design (CAD) and computer assisted manufacturing (CAM), especially in the mould and die sub-sector, to introduce, train, and provide facilities to the private sector, especially to manufacturers using computers.
7. Plans to set up technology transfer centres to gather technological information, study new technologies, and modify them for local use.

A technology park was established in 1988 in Bandar Tun Razak, Cheras, Kuala Lumpur which will act as a mechanism to stimulate the growth of high-tech industries in the country. The park is expected to provide a focus for a series of programmes designed to stimulate, promote, support, and commercialize innovative concepts drawn from R & D in local research institutes and universities. If this proves successful, it will become a model for similar facilities which can be developed in other urban centres. In this regard, there is a need to emphasize the importance of what has been termed as the 'breeding and nurturing of innovation clusters'. Much of the dynamic effects of innovation may come not from the intrinsic value of one isolated innovation but from the way many innovations meet and join into a cluster of innovations. In this respect, it may be added that domestic manufacturing enterprises still have a long way to go. According to Debresson (1989: 1–16):

> A development policy aimed at innovation must think beyond individual innovative capabilities in terms of technological systems and innovations clusters. Relying on market forces and those economic inducements which most favour the meeting of different innovations, an intelligent technical policy can maximise their effects. This can be done by fostering interaction by potential investors.

### *Special R & D Allocation under IRPA*

IRPA (Intensification of Research in Priority Areas) was instituted in early 1987 to monitor and approve research projects when a special R & D fund was created under the Fifth Malaysia Plan (1986–90). While its principal function is to determine the priority areas for research at the national and institutional levels as well as to evaluate research proposals based on these priorities, and thereafter monitor their progress, it is also responsible for integrating strategic planning in S & T with national development planning and providing a platform for interaction between the scientific community and planners/decision-makers. Lastly, it is also responsible for ensuring the development of linkages between public sector users of R & D and R & D implementation agencies.

IRPA was created with the view of instituting a more centralized and co-ordinated management system for R & D and technology development within the country given the manpower and budgetary constraints and the need to achieve effective and efficient S & T application and utilization. An important outcome of this mechanism is that it has initiated the process of consultation and consensus-building among the relevant people from both the public and private sectors (Malaysia, 1991b: 188).

### *Constraints in S & T Policy Implementation*

Given the socio-economic environment within which the S & T institutional framework operates, there are a number of areas which could have been improved upon, especially in terms of the implementation of strategies. In this sense, Malaysia has missed a number of opportunities. Without redressing these areas, the effectiveness of any future S & T programme will not be realized as envisaged by policy-makers, and this will certainly have a stifling impact on the pace of industrialization in the country.

Although a policy-making structure for S & T had already been instituted prior to the launching of the Action Plan for Industrial Technology Development, in practice it was not exactly clear which organization had the final say over S & T programmes and projects. In view of the proliferation of S & T related and R & D agencies, decision-making tended to be fairly diffused, apart from the additional problem of co-ordination. In the absence of a national programme for S & T development previously, the various agencies carried out their activities without the benefit of a national S & T framework. Thus, according to Ahmad Zaharudin Idrus (1988: 91–7):

> Failures of R & D are often associated with the lack of infrastructure to carry through the development of appropriate trials for research results. There are insufficient funds to test new ideas or results on the semi-commercial scale. There is no capability for testing products in the market. Research is often undertaken in isolation, only to solve few problems. The programme of research is not carried out with a complete understanding of its mission.

The extent of the NCSRD's effectiveness in promoting S & T development was ambiguous during its inception, while its capacity was solely advisory in nature. In addition, it lacked the manpower and resources to carry out its responsibilities. The poor integration of local capabilities with local needs stemmed from poor assessment of needs and of ways to satisfy those needs. Apart from the lack of expertise, in general, public sector institutes that conducted basic research faced real problems in ensuring their relevance to the needs of industry.

An important component of technology transfer relates to the commercialization of research, involving the transfer of knowledge regarding research results from the research laboratory to the market-place. This aspect is becoming even more important in view of the accelerating pace with which S & T is advancing and the reduction in time between basic advances and their being incorporated in products or processes. In the industrial countries, a wide variety of mechanisms have been initiated with a great deal of success, including incubators, innovation centres, science or technology parks, research clubs, selective dissemination of information, and collaborative programmes between research institutes or universities and industry.

In Malaysia, very little effort has so far been made to facilitate and promote the commercialization of R & D results. This arises not only because of the lack of a R & D tradition within the public sector research institutes, but also because of the relatively few linkages developed between these institutes and the universities and industry. While the mechanisms to promote such linkages are still to be fully developed, there is also a problem of perceptions on all sides on the ability of each to create a permanent kind of interaction. The research institutes are perceived by industry to be lacking in market-orientation while the latter is perceived by the former to be disinterested in forging the necessary linkages. In addition, there is a tendency for industry, especially the larger enterprises, to undertake its own research while the small-scale enterprises are unlikely to do any research. With the exception of agriculture-based industry, there is no organizational intermediary which bridges the gap between the R & D institutions and the private sector.

Associated with the above constraint is the lack of funding support or venture capital for such commercialization activities. Existing loan facilities or financial support are mainly meant to satisfy the broad objectives of stimulating industrial development and export promotion. In order to stimulate more R & D efforts, which should also materialize into tangible commercial concerns, there must be a funding mechanism that will allow R & D activities to flourish, particularly within the universities or research institutes. The diversity of constraints faced by the manufacturing sector as it seeks to adjust to the changing global competitive environment are to a significant degree a reflection of its growth patterns and past industrial strategies. These constraints will define the sort of strategic options from which the country may be able to choose as it seeks to advance its industrialization process.

Public sector research institutes are financed fully by the state except for PORIM and RRIM whose activities are financed by a cess on palm oil produced and rubber exported respectively. Unlike these two institutes, budgetary allocations for the other research institutes are on an annual basis and are usually for operating and fixed expenses with major proportions of the budget earmarked for emoluments and for acquiring equipment which does not fall into the category of development expenses. There is no budget for financing of specific projects but the funds to continue projects started are usually incorporated into the main annual budget of the different institutes (Asian and Pacific Centre for Transfer of Technology, 1986: 172). As far as development, expansion, and construction of scientific and technological infrastructure are concerned, the allocations are made available to the various research institutes through a lump-sum allocation for development expenditure at the beginning of a Five-Year Development Plan where such projects have been submitted to be included as projects under the Plan. This kind of budgetary allocation mechanism may become irrelevant in the future given that these research institutes have to respond positively and quickly to any external changes. To overcome this, research institutes ought to be self-financing, whereby they have to take a more explicit commercial stance to enhance their potential contribution to S & T development (Malaysia, 1991b: 205).

# 7
# Explicit Priorities in Technology Development

## Reappraisal of S & T Strategies

THE state has expressed its intent to give more emphasis to industrial R & D, although this has not been fully substantiated by any sense of urgency in many areas. According to the Mid-Term Review of the Fifth Malaysia Plan (1986–90):

> Increased importance was accorded towards the development of research and development support for the manufacturing sector. Measures included the provision of sufficient research funds, the strengthening of R & D institutions and development of indigenous technological capability. Of the total allocation for R & D during the Fifth Plan amounting to $400 million, about 35 percent was allocated for industrial R & D. The purpose was to encourage innovations and advances in the technological base, with a view to enhance the effectiveness of institutions and universities in undertaking and stimulating the use and application of industrial research for industry. Efforts were undertaken to restructure R & D institutions with the view to streamline and strengthen these institutions to meet the technology demands of the manufacturing sector. Simultaneously, fiscal incentives such as double deduction on expenditure incurred in approved R & D, were made available in the new incentive package, in order to promote private sector involvement in R & D.
>
> High priority was given to the development of indigenous technological capability to absorb, adapt, and modify existing and imported technology as well as the development of new indigenous technology. Measures towards achieving these objectives were implemented through restructuring and the reorientation of the education curriculum at tertiary institutions. Courses in electrical, electronics, mechanical, and chemical engineering were emphasized to ultimately produce high level manpower with the necessary knowledge and skills required by the industries. In addition, students with research inclination studying abroad were encouraged to further pursue their studies. They were also encouraged to acquire as much practical working experience and skills as possible through attachments in foreign companies (Malaysia, 1989a: 191–3).

In 1982, Malaysia spent about $295 million or 0.5 per cent of its gross domestic product on R & D. Although the percentage had increased slightly to 0.8 per cent of GDP in 1988, this was still very low

TABLE 7.1
R & D Expenditure in Selected Countries (as percentage of GDP)

| *Country* | *Public Sector* | *Private Sector* | *Total* |
|---|---|---|---|
| United States (1985) | 1.4 | 1.3 | 2.7 |
| Japan (1985) | 1.3 | 1.3 | 2.6 |
| West Germany (1984) | 0.8 | 1.6 | 2.4 |
| Switzerland (1981) | 0.5 | 1.9 | 2.4 |
| Sweden (1981) | 0.6 | 0.7 | 1.3 |
| South Korea (1988) | 0.8 | 1.2 | 2.0 |
| Taiwan (1984) | 0.6 | 0.3 | 0.9 |
| Malaysia (1988) | 0.7 | 0.1 | 0.8 |

*Source*: Bureau of National Economic Policy Studies, Institute of Strategic and International Studies, Kuala Lumpur.

given that 1 per cent of the GDP is generally considered to be the minimum level at which R & D can begin to effectively provide the critical mass and support for socio-economic development in the country. The national ratio also compared unfavourably with those of the NICs, such as South Korea, which spent 2.0 per cent of its GDP on R & D in 1988, and the industrial countries, such as Japan and the United States, which spent 2.6 per cent and 2.7 per cent of their respective GDPs in 1985 (Table 7.1). Of the national R & D expenditure, the private sector, despite its supposedly vital role in the industrialization process, contributed only about 10 per cent compared with 45 per cent and 70 per cent in the case of South Korea and Japan respectively (Malaysia, 1986a: 262).

While it is difficult to establish the relationship between the level of R & D expenditure and technology development (which also depends on many other factors in addition to R & D allocations), one may conclude that the development of an indigenous technological capability contributes positively to higher productivity, and such capability can only be effected through strategic decisions on the future course of S & T development. If Malaysia were to narrow the technological gap between herself and the industrial countries, it must be willing to reappraise its existing policies on technology development in order to map out its strategic options for the 1990s and beyond. Any extensive appraisal in this field must require an examination of three critical areas: human resource development (HRD), reducing technology dependence and upgrading domestic technological capability, and the creation of a conducive S & T environment.

## Human Resources for Technology Development

The development of human resources within a country is probably the most critical function of the state in terms of long-run technological enhancement. Countries which do not invest appropriately in upgrading their educational levels as well as the skills of their industrial labour force will inevitably suffer a relative economic decline in the future. Therefore, appropriate focus needs to be given to HRD so that the economy can anticipate potential skill bottlenecks and adjust educational and training programmes accordingly.

As a critical component of HRD, educational planning has to concentrate on the development of the curriculum at all levels, and to appraise continually its contribution towards the changing needs of the new industrial structure. The development of a curriculum must also consider its impact on individual and collective attitudes towards 'blue-collar' jobs. Unless values inculcated in schools with regard to employment of a technical nature are correct, there may arise the problem of a divergence between the supply and actual demand for this type of labour. Adjusting values may also include a reorientation of work ethics or discipline which must cater for a changing work environment of an industrial nature. Priority should therefore be placed on changing the curriculum, particularly at the secondary level, from being too academic to more vocation orientated (Anuwar Ali et al., 1979: 283–310). Such a strategy would benefit the majority of students since the intake into higher institutions of learning is relatively limited.

During the last few years, this issue has been met with a heightened sense of urgency on the part of the planning machinery. As noted by the Sixth Malaysia Plan (1991–5), a curriculum reform was started in 1989:

> a significant change included in the Integrated Secondary School Curriculum (KBSM) was the greater emphasis given to business-related and pre-vocational subjects. At the lower secondary level, a new subject called 'Living Skills', incorporating business knowledge was introduced with the objective of exposing students to aspects of technology, commerce and entrepreneurship (Malaysia, 1991b: 161).

Apart from this, the Plan also intends to increase the intake into public skill training institutions through the completion of 8 new secondary vocational schools, the expansion of 29 existing vocational schools, and other public training institutions (Malaysia, 1991b: 177).

In spite of the above, the existing 'educational approach' seems to confine the planner's attention to high-level manpower needed in the modern sector, that is, urban-based employment. The bias towards high-level education, particularly degree programmes, has been explicitly acknowledged in the Fourth Malaysia Plan so that there has been significant shortages of middle-level manpower, especially in the scientific and technical fields (Malaysia, 1981a: 92). Given that planners tend to emphasize high-level manpower, it is important to note that a substantial proportion of such manpower is engaged in the public sector rather than in industry. In 1980, approximately 76 per cent of professional and technical personnel were employed in government services, and it was then projected that the public sector would absorb 73 per cent of the increase during the 1980–5 period (Malaysia, 1981b: 58).

TABLE 7.2
R & D Manpower in Public and Private Sectors, 1989

| *Qualification* | *Public (%)* | *Private (%)* | *Total (%)* |
|---|---|---|---|
| Ph.D. | 1,254 (97) | 35 (3) | 1,289 (100) |
| Masters | 2,181 (97) | 66 (3) | 2,247 (100) |
| B.Sc. | 1,679 (84) | 322 (16) | 2,001 (100) |
| Subtotal | 5,114 (92) | 423 (8) | 5,537 (100) |
| Others[a] | 6,656 (82) | 1,412 (18) | 8,068 (100) |
| Total | 11,770 (87) | 1,835 (13) | 13,605 (100) |

*Source*: Malaysia (1991b).
[a]Includes sub-professionals, technicians, and other supporting staff.

According to the Sixth Malaysia Plan (1991–5), 'despite the high rate of expansion in the output of high- and middle-level S & T related manpower from institutions of higher learning, the imbalance in the type and number of manpower produced and required by the nation continued to be a problem' (Malaysia, 1991b: 195). Table 7.2 indicates that a total of 13,605 personnel were involved in R & D activities in 1989, out of which 5,537 (41 per cent) were research scientists and the rest were supporting staff. With 87 per cent of the total R & D personnel, the public sector represented the largest source of R & D manpower; a substantial number of them were engaged in agriculture-related research.

At the same time, a large proportion of R & D personnel in the public sector were in either basic or upstream research compared to applied or developmental research, as a consequence of which there was minimal impact on industry-related R & D. On the other hand, the number of R & D personnel in the private sector was too small to produce any significant impact on indigenous market-driven R & D (Table 7.3). When comparisons were made with other countries in terms of the ratio of research scientists to the total population, the ratio for Malaysia (at 400 per million population) was indeed very low, while the equivalent ratio for Japan was 6,500 per million; United Kingdom 3,200 per million; West Germany 3,000 per million; and South Korea 1,300 per million (Malaysia, 1991b: 197).

Educational planning seems to neglect not only middle-level manpower but more importantly the requirements of the vast majority of the country's labour force—mostly the semi-skilled and unskilled workers in the urban centres and the majority of rural workers. Policy formulation must consider that in most industries the proportion of unskilled labour is still substantial (Malaysia, 1981b: 74–7). This was particularly conspicuous in the electronics industry where, in 1980, 81.9 per cent of the labour force was considered unskilled. In 1986, according to a special

TABLE 7.3
R & D Manpower by Specialization and Qualification, 1989 (number)

| | *Ph.D.* | | *Masters* | | *B.Sc.* | | |
|---|---|---|---|---|---|---|---|
| *Specialization* | *Public* | *Private* | *Public* | *Private* | *Public* | *Private* | *Total* |
| Engineering | 142 | 7 | 486 | 18 | 585 | 115 | 1,353 |
| Computer | 33 | 0 | 138 | 1 | 40 | 30 | 242 |
| Medical | 87 | 0 | 134 | 0 | 112 | 0 | 333 |
| Agriculture | 347 | 19 | 528 | 15 | 210 | 43 | 1,162 |
| Basic | 491 | 9 | 484 | 32 | 482 | 134 | 1,632 |
| Others | 154 | 0 | 411 | 0 | 250 | 0 | 815 |
| Subtotal | 1,254 | 35 | 2,181 | 66 | 1,679 | 322 | 5,537 |
| Total | 1,289 | | 2,247 | | 2,001 | | |

*Source*: Malaysia (1991b).

survey on the electrical industry by the Ministry of Labour, production workers formed the largest group at 72.1 per cent of the total labour force in the industry; this group was composed of almost entirely semi-skilled and unskilled workers (Malaysia, 1988a: 76).

An important consideration is that manpower is not only a major factor of production, but must also be seen as the major beneficiary of the effects of industrialization and economic development. According to the Ministry of Labour and Manpower:

> ... the provision of gainful employment has been regarded as one of the most powerful poverty eradication tools in development plans in Malaysia. The unemployed and underemployed are not only likely to be poor but also ill-equipped to participate in the economic life of the nation. The thrust of the national development effort has been to create more employment opportunities, to upgrade skills and levels of income so as to improve the quality of life in Malaysia (Malaysia, 1981b: 25).

Thus, the development of human resources through the education system is an important prerequisite for economic development and a good investment of scarce resources, provided the pattern and quality of educational output is geared to the economy's manpower needs.

The implementation of a HRD programme must encompass not only the elements of technological progress through educational planning and the upgrading of existing skills but also the basic needs of the labour force. Improvement in these areas are important because they are related to 'investment in human capital'. Such investments may take many forms, including public sector allocations for health facilities, low-cost housing subsidies, on-the-job training and retraining, formally organized education, study programmes, and adult education. Investment in human capital can overcome the impediments to greater productivity, such as poor health, illiteracy, unresponsiveness to new knowledge, a lack of incentive, and immobility.

In general, therefore, HRD involves two crucial and related components. First, it encompasses the need for an evaluation of existing manpower and an assessment of future skill requirements in all sectors of the economy. In the Malaysian context in the early 1990s, the demand for skilled industrial labour is extremely crucial. Such evaluation and assessment, requiring an effective institutional framework, would allow the planning authorities to formulate the numerous measures needed to ensure that resources are available when required. While these capabilities are within the realm of the public sector, the interaction between it and industry is equally important. Second, at the micro level HRD must be concerned with the priorities and commitment of the top management in industry with respect to the development of specific skills related to technological changes as well as facilities to ensure a conducive environment for work and upward mobility (Saxena, 1990).

Although education has received substantial public expenditure allocation, what seems to be lacking is an appropriate institutional framework which will cater to the needs of the existing industrial labour force, particularly the need to upgrade skills, acquire new skills, and instil an innovative mind at the managerial and technical levels. This lack has to be remedied through educational and manpower planning. Given the resource constraint, the issue is whether a substantial increase in public expenditure can be allocated for the enhancement of technical skills amongst the existing labour force, including the majority of school-leavers, given that the intake into institutions of higher learning is limited.

However, education and training has to compete for public expenditure allocations with other sectors such as the economic (including agriculture, mineral resource development, commerce and industry, transport, communications, and energy and public utilities), social (besides education and training, this includes health, housing, culture, youth, sports, local councils and welfare services, community development), security, and public administration sectors. Out of the total estimated expenditure (amounting to $74,063 million) during the Fourth Malaysia Plan (1981–5), the allocation for education and training accounted for 6.3 per cent compared to 10.1 per cent for security. Table 7.4, however, indicates that education and training since then has been accorded more emphasis so that its share in the total public development allocation and expenditure has increased substantially. Despite the absolute reduction in public expenditure during the 1986–90 period as a consequence of the 1985–6 economic recession, the allocation and expenditure given to education and training increased in absolute terms, and more so in relative terms. The allocation for the Sixth Plan period is similarly high.

Table 7.5 shows that student enrolment in post-secondary education

TABLE 7.4
Public Development Allocation and Expenditure by Sector, 1981–1995 ($ million)

| *Sector* | *Fourth Plan Allocation Expenditure 1981–5* | *Estimated Expenditure 1981–5* | *Fifth Plan Revised Allocation 1986–90* | *Estimated Expenditure 1986–90* | *Sixth Plan Allocation 1991–5* |
|---|---|---|---|---|---|
| Economic | 30,103.81 | 55,777.63 | 24,048 | 22,886 | 31,236 |
| | (61.41) | (75.31) | (64.5) | (64.8) | (56.8) |
| Social | 10,340.91 | 9,980.19 | 9,046 | 8,764 | 13,468 |
| | (21.09) | (13.48) | (24.3) | (24.8) | (24.5) |
| Education and training | 4,840.09 | 4,687.59 | 5,812 | 5,700 | 8,501 |
| | (9.87) | (6.33) | (15.6) | (16.1) | (15.5) |
| Security | 7,741.78 | 7,494.58 | 2,955 | 2,527 | 8,408 |
| | (15.79) | (10.12) | (7.9) | (7.2) | (15.3) |
| Administration | 838.84 | 810.60 | 1,241 | 1,123 | 1,888 |
| | (1.71) | (1.09) | (3.3) | (3.2) | (3.4) |
| Total | 49,025.34 | 74,063.00 | 37,290 | 35,300 | 55,000 |
| | (100) | (100) | (100) | (100) | (100) |

*Sources*: Malaysia (1986a, 1991b).
*Notes*: Education and training is categorized under Social Allocation and Expenditure and is therefore not included (again) in the total. Figures in parentheses are percentages of the total.

(including teacher training, certificate/diploma level, and degree level) was 1.3 per cent, 2.5 per cent, 3.7 per cent, and 4.6 per cent during 1970, 1980, 1985, and 1990 respectively, and in 1995, the percentage is expected to increase to 5.1 per cent. With the exception of the enrolment at the certificate/diploma level, the enrolment at the degree level recorded the highest increase at 72.4 per cent during the 1981–5 period and 58.6 per cent during the 1986–90 period. Increasingly, the government is subjected to tremendous political and social pressures to increase the student intake at the degree level. Although the increase in student enrolment at the post-secondary level has been commendable during the 1980s, a substantial proportion of the overall student intake is still concentrated at the primary and secondary levels.

During the 1981–5 period, however, total enrolment in local degree and diploma courses increased by 89 and 104 per cent respectively, although the percentage increase during the 1986–90 period in both categories declined to 59 and 12 per cent respectively (Table 7.6). Although the total enrolment in local degree courses is not projected to increase as fast during the 1991–6 period, an interesting trend is that the projected increase in the science and technical courses is relatively substantial. A similar trend is expected to be duplicated for the diploma courses intake during the 1991–6 period.

TABLE 7.5

Student Enrolment by Level of Education, 1970–1995 (percentage)

| | | | | | | Increase (%) | |
|---|---|---|---|---|---|---|---|
| *Educational Level* | *1970* | *1980* | *1985* | *1990* | *1995* | *1981–5* | *1986–90* |
| Primary[a] | 75.0 | 63.8 | 62.8 | 63.7 | 61.5 | 9.1 | 13.8 |
| Lower secondary | 19.3 | 25.8 | 24.6 | 22.8 | 21.9 | 13.1 | 2.4 |
| Upper secondary[b] | 4.4 | 7.9 | 8.9 | 8.9 | 11.5 | 34.4 | 10.6 |
| Post-secondary[c] | 0.6 | 0.9 | 1.4 | 1.8 | 1.6 | 58.2 | 43.4 |
| Teacher education | 0.1 | 0.4 | 0.4 | 0.5 | 0.5 | 25.0 | 30.3 |
| Certificate/diploma level | 0.3 | 0.5 | 0.9 | 0.9 | 1.3 | 143.6 | 22.6 |
| Degree level | 0.3 | 0.7 | 1.0 | 1.4 | 1.7 | 72.4 | 58.6 |
| Total | 100.0 | 100.0 | 100.0 | 100.0 | 100.0 | 13.7 | 10.5 |

*Sources*: Malaysia (1981a, 1986a, 1991b).

*Note*: Total Student Enrolment:

1970—2,240,064
1980—3,150,095
1985—3,748,650
1990—4,142,380
1995—5,148,600

[a]For 1985 onwards, primary level enrolment includes pre-school enrolment.

[b]Upper secondary includes technical and vocational training.

[c]Post-secondary includes Kolej Tunku Abdul Rahman.

TABLE 7.6
Enrolment of Degree and Diploma Holders in Local Institutions, 1980–1995

| *Course* | *1980* | *1985* | *1990* | *1995* | *Increase (%)* | | |
|---|---|---|---|---|---|---|---|
| | | | | | *1981–5* | *1986–90* | *1991–5* |
| Degree | | | | | | | |
| Arts | 9,727 | 20,350 | 34,660 | 51,410 | 109 | 70 | 48 |
| | (49) | (54) | (58) | (57) | | | |
| Science | 8,046 | 12,330 | 16,450 | 24,210 | 53 | 33 | 47 |
| | (40) | (32) | (27) | (27) | | | |
| Technical | 2,245 | 5,160 | 8,920 | 14,050 | 130 | 73 | 58 |
| | (11) | (14) | (15) | (16) | | | |
| Total | 20,018 | 37,840 | 60,030 | 89,670 | 89 | 59 | 49 |
| | (100) | (100) | (100) | (100) | | | |
| Diploma | | | | | | | |
| Arts | 5,063 | 12,830 | 14,920 | 20,620 | 153 | 16 | 38 |
| | (41) | (51) | (53) | (47) | | | |
| Science | 3,279 | 5,440 | 4,750 | 8,860 | 66 | –13 | 87 |
| | (27) | (22) | (17) | (20) | | | |
| Technical | 3,920 | 6,800 | 8,340 | 14,740 | 73 | 23 | 77 |
| | (32) | (27) | (30) | (33) | | | |
| Total | 12,262 | 25,070 | 28,010 | 44,220 | 104 | 12 | 58 |
| | (100) | (100) | (100) | (100) | | | |

*Sources*: Malaysia (1986a, 1991b).
*Note*: Figures in parentheses are percentages of the total.

Although high rates of enrolment were sustained for science and technical courses in line with efforts to increase the output of high- and middle-level manpower in these fields (the output ratio of arts to science and technical courses improved from 52 : 48 in 1980 to 50 : 50 in 1985), the output of arts and humanities graduates continued to remain at a high level to cater for the demand for secondary school teachers as well as administrative and managerial personnel (Malaysia, 1986a: 149). Table 7.7 shows that during the 1981–5 period, 34.7 per cent of graduates were in science while 10.1 per cent were from the technical stream compared to 55.2 per cent from the arts and humanities. There was no drastic change at the degree level during the 1986–90 period; the percentage output declined marginally in the arts courses to 52.6 per cent, and in the science courses to 33.1 per cent.

For the whole of the Sixth Malaysia Plan period (1991–5), the situation is projected to remain almost unchanged with the proportion of arts graduates increasing to 60.7 per cent, while the ratios of science and technical graduates will fall to 25.5 and 13.8 per cent respectively. Nevertheless, there will be an absolute increase in graduates in these two areas. Although the output of diploma holders is higher in the science and technical courses, the output from the arts courses is still high with 42.9 per cent during the 1981–5 period and 50.1 per cent during the

TABLE 7.7
Output of Degree and Diploma Holders, 1981–1995

| *Course* | *1981–5* | *1986–90* | *1991–5* |
|---|---|---|---|
| Degree | | | |
| Arts | 14,802 | 27,780 | 50,250 |
| | (55.2) | (52.6) | (60.7) |
| Science | 9,317 | 17,510 | 21,110 |
| | (34.7) | (33.1) | (25.5) |
| Technical | 2,719 | 7,550 | 11,430 |
| | (10.1) | (14.3) | (13.8) |
| Total | 26,838 | 52,840 | 82,790 |
| | (100.0) | (100.0) | (100.0) |
| Diploma | | | |
| Arts | 9,808 | 18,450 | 29,970 |
| | (42.9) | (50.1) | (53.3) |
| Science | 5,636 | 7,950 | 10,500 |
| | (24.7) | (21.6) | (18.7) |
| Technical | 7,404 | 10,450 | 15,750 |
| | (32.4) | (28.3) | (28.0) |
| Total | 22,848 | 36,850 | 56,220 |
| | (100.0) | (100.0) | (100.0) |

*Sources*: Malaysia (1986a, 1991b).

TABLE 7.8

Output of Skilled and Semi-skilled Manpower from Public Training Institutions, 1981–1995

| | | | | | | | *Increase (%)* | |
|---|---|---|---|---|---|---|---|---|
| *Courses* | *1981–5* | *(%)* | *1986–90* | *(%)* | *1991–5* | *(%)* | *1986–90* | *1991–5* |
| Engineering trades | 47,091 | 59.1 | 64,040 | 57.7 | 132,760 | 67.4 | 36.0 | 107.3 |
| Building trades | 8,162 | 10.3 | 12,850 | 11.6 | 16,750 | 8.5 | 57.4 | 30.4 |
| Printing trades | 460 | 0.6 | 170 | 0.1 | 550 | 0.3 | –63.0 | 223.5 |
| Commerce | 9,230 | 11.6 | 8,890 | 8.0 | 20,120 | 10.2 | –3.7 | 126.3 |
| Agriculture | 4,459 | 5.6 | 4,470 | 4.0 | 1,120 | 0.6 | 0.2 | –74.9 |
| Home Science | 5,847 | 7.3 | 4,690 | 4.2 | 9,170 | 4.6 | –19.8 | 95.5 |
| Others | 2,954 | 3.7 | 11,190 | 10.1 | 6,500 | 3.3 | 278.8 | –41.9 |
| Skill upgrading | 1,437 | 1.8 | 4,730 | 4.3 | 10,000 | 5.1 | 229.2 | 111.4 |
| Total | 79,640 | 100.0 | 111,030 | 100.0 | 196,970 | 100.0 | 39.4 | 77.4 |

*Sources*: Malaysia (1986a, 1991b).

1986–90 period. It is of interest to note that the output of diploma holders from the technical courses is to be sustained at a relatively high level, that is, 28.0 per cent of the total diploma holders during the Sixth Malaysia Plan period (1991–5), although this is lower than that in the 1981–5 period.

To complement the output of science and technical personnel at the degree and diploma levels, the output of skilled and semi-skilled manpower from the vocational and technical schools and industrial training institutes is equally important. Table 7.8 shows that more than half the output from these public training institutes was from engineering; with 59.1 per cent of the total during the 1981–5 period and 57.7 per cent during the 1986–90 period. During the 1991–5 period, the percentage is expected to increase to 67.4 per cent; indeed, a reflection of the recognition by planners of the shortage of skilled labour. In this regard, the IMP has proposed that the number of engineers/technicians in the manufacturing sector be increased from the 12,000 in 1986 to at least 50,000 by 1995, emphasizing areas such as textiles, electronics, and mechanical, metallurgical, chemical, and industrial engineering (UNDP/UNIDO, 1985: 258–9).

If one examines the public expenditure allocations for education and training programmes, as given in Table 7.9, one finds that a substantial proportion has been allocated for primary and secondary education (although with a declining share of 37.9 per cent, 32.5 per cent, and 31.6 per cent during the 1981–5, 1986–90, and 1991–5 periods respectively), while allocations for training programmes for technical, vocational, or industrial training accounted for less than 20 per cent of the total educational budget during each of the three Plan periods. This reflects the emphasis on academic and literary achievement at the primary and secondary levels given a population structure that is youthful in character.

According to Semudram and Mokhtar Tamin (1989), programmes at the vocational and technical levels have been a weak link, and this top heavy structure has resulted in the emergence of high unemployment among graduates during the mid-1980s. This mismatch has resulted in stiff competition for jobs at the lower end between degree and diploma holders. As such, the phenomenon of graduate unemployment has often been attributed to the weakness of the education and training systems (Zainal Abidin Abdul Rashid, 1990: 209–19).

In the case of South Korea, on the other hand, its education and training programmes are based on the need to achieve an appropriate balance so that over-production at the secondary and tertiary levels is avoided. Since the reduction of educated unemployment became a major objective for South Korea, emphasis was placed on vocational and technical education with the aim of attaining a 70 : 30 enrolment ratio between vocational/technical and high school by 1980. The preparation of South Korean high school graduates for the job market has become a critical issue, because the effective supply of graduates from colleges or

TABLE 7.9
Public Expenditure Allocations for Education and Training Programmes, 1981–1995

| Programmes | *Estimated Expenditure 1981–5 (A)* | | *Estimated Expenditure 1986–90 (B)* | | *Expenditure Allocation 1991–5 (C)* | | *Expenditure Increase (%)* | |
|---|---|---|---|---|---|---|---|---|
| | *($ million)* | *(%)* | *($ million)* | *(%)* | *($ million)* | *(%)* | *(A–B)* | *(B–C)* |
| Primary education | 759 | 15.5 | 760 | 13.3 | 1,160 | 13.7 | 0.1 | 52.6 |
| Secondary education | 1,094 | 22.4 | 1,091 | 19.2 | 1,523 | 17.9 | –0.3 | 39.6 |
| Technical and vocational schools | 288 | 5.9 | 452 | 7.9 | 480 | 5.6 | 56.9 | 6.2 |
| Polytechnics | 123 | 2.5 | 228 | 4.0 | 296 | 3.5 | 85.4 | 29.8 |
| Higher education[a] | 1,934 | 39.6 | 1,499 | 26.3 | 2,295 | 27.0 | –22.5 | 53.1 |
| Teacher training | 149 | 3.0 | 144 | 2.5 | 334 | 3.9 | –3.4 | 131.9 |
| Other educational support programmes | 135 | 2.8 | 1,208 | 21.2 | 1,636 | 19.3 | 794.8 | 35.4 |
| Industrial training[b] | 265 | 5.4 | 299 | 5.2 | 610 | 7.2 | 12.8 | 104.0 |
| Commercial training[c] | 1 | – | 7 | 0.1 | 27 | 0.3 | 600.0 | 285.7 |
| Management training[d] | 141 | 2.9 | 12 | 0.2 | 140 | 1.6 | –91.5 | 1,066.7 |
| | 4,889 | 100.0 | 5,700 | 100.0 | 8,501 | 100.0 | 16.6 | 49.1 |

*Sources*: Malaysia (1986a, 1991b).
[a]All local universities, Kolej Tunku Abdul Rahman, and Institut Teknologi MARA.
[b]Industrial Training Institutes, MARA Vocational Institutes, and Youth Training Centres.
[c]MARA Commercial Institutes and Youth Entrepreneurship Institute.
[d]National Institute of Public Administration (INTAN) and Institut Aminuddin Baki.

universities is constrained by the government's inability to expand the enrolment in colleges or universities, and also because industry's efforts to replace highly paid college or university graduates with lower-paid high school graduates is expected to accelerate as the wage differentials between the two groups widen (Lee Kye-Woo, 1987: 28–54).

More significant in the context of Malaysian industrial development is the increase in expenditure allocation for technical and vocational training during both the 1986–90 and 1991–5 periods by 56.9 and 6.2 per cent respectively. This is also reflected in the allocations for the polytechnics and industrial training. As indicated in Table 7.9, university education has also been given priority during the 1980s with substantial increases in expenditure allocations. Although the universities (including Malaysians trained in overseas universities) have so far produced a sufficient number of graduates in the pure sciences, their opportunities for useful employment are generally restricted to teaching, given the limited nature of R & D within domestic industries. But the increase in national emphasis and commitment towards R & D demands that there be an adequate pool of scientific and technical manpower capable of acquiring, adapting, and assimilating imported technology and, finally, generating indigenous technology. This role must be effectively filled by the local universities.

While the state undertakes to increase the enrolment of postgraduate students to meet national needs for high-level manpower, their effective contribution in undertaking research and in effecting technological change must also be nurtured within the private sector through their direct role in R & D activities, either by in-house research or through collaborative research activities with the universities themselves or public sector research institutes. However, as noted earlier, the potential for such collaborative research activities has not been fully explored, and if there is such collaboration, it is still in its infancy. A contributing factor in this situation, according to the Fifth Malaysia Plan, is that the universities are hampered by constraints in funding for both teaching and research, by a shortage of experienced staff, and by low enrolment in graduate studies in the science and engineering disciplines (Malaysia, 1986a: 261–2).

As indicated earlier, HRD for industry is not necessarily confined to programmes within the education system, but must also be seen in a wider context in which the existing industrial labour force must be given appropriate training for skill development. In this context, the private sector too, more specifically the larger manufacturing enterprises, must play a significant role in enhancing the skills of the industrial labour force in return for the 'active promotion and the maintenance of a favourable investment climate by the government' (Malaysia, 1981a: 133). Since the private sector is entrusted with the role of spearheading the industrialization drive, it should be responsible for providing comprehensive training programmes for its labour force, including training with technology suppliers and joint-venture partners, and on-the-job training on the factory floor by engaging consultants or experts in product and process development.

However, since skill training is costly, most domestic manufacturing enterprises are reluctant to allot funds for such training. This is because they are either too small to afford the time and funds to train their workers or they fear that their employees, once trained, will be attracted to the larger enterprises. Apart from these factors, there is also the constraint associated with the lack of a proper mechanism to ensure that manufacturing enterprises upgrade the skills of their work-force.

If the Japanese experience is to be duplicated, domestic industries should be encouraged to establish 'factory schools' in order to increase basic technical competence (Hayashi, 1983). However, it must be noted that such a programme may only be possible when the country has achieved a reasonably high level of industrialization in which there already exists numerous large-scale manufacturing firms in a number of industrial sub-sectors which can sustain the training costs and provide experienced and highly skilled personnel to train the other employees. In the early 1990s, the size of domestic manufacturing enterprises in Malaysia is relatively small by industrial country standards, while in the case of the larger ones, they are mostly owned or managed by foreign interests which are not particularly committed to providing the necessary facilities to upgrade the skills of their labour force.

The increasing importance of manufacturing in terms of national output and employment creation during the 1990s and beyond gives rise to the need for preparing the working population with the necessary scientific, technical, and managerial skills. This calls for well-conceived training programmes at the firm level, particularly so in terms of the country's aim to develop not only heavy and high-technology industries but also industries that will be able to compete in international markets. However, decision-making at the industry level, given the existing structure of the manufacturing sector as a whole (including the role of foreign MNCs and the incentives system), seems to place little emphasis on labour's capacity to innovate, such capability being an important element of technical progress. The situation is further complicated by the comparatively small domestic industrial base whose capacity in terms of producing intermediate and capital goods is still negligible, hence perpetuating the state of dependence on imported technologies. Although technological diffusion may be externally induced, it is important for the expansion of the country's industrial base that the domestic technological capability be enhanced as quickly as possible.

It is, of course, essential to strike a balance between the need to maintain reasonable standards in education and the need to upgrade labour's technical skills because of constraints imposed upon public expenditure allocations. In this respect, economic and educational planners must have a proper perspective on these needs. In view of the country's drive towards heavy industries and high-technology industries, it is appropriate that a comprehensive programme for upgrading existing skills and acquiring new skills be formulated. Such a programme must also consider issues relating to technology transfer from the industrial countries, the adaptive skills of the existing labour force, and R & D on appropriate techniques, processes, and products.

The development of human resources for technological enhancement must be regarded as a cumulative process involving the expansion and consolidation of efforts by the state to enhance all kinds of industrial skills, with the private sector acting in tandem. These efforts must be complemented by a manpower development programme which has to be maintained at a sufficiently high level through constant training, a conducive environment for R & D, and incentives and awards to attract and retain trained and experienced personnel in their appropriate fields.

While the critical need for various skills in industry is recognized, increasing efforts must also be made to build expertise at the management level where decision-making takes place and at the floor level where decisions are implemented. In the manufacturing sector, there already exists an imbalance in terms of both the lack of engineering personnel and the inappropriate proportion between this group and the technical personnel. The inadequacy of the education system has been a factor in this imbalance, although one can also argue that other factors such as the industrial wage structure, the social recognition system in society, and the structure of domestic industries have had an equally important bearing on this matter.

According to the Ministry of Labour and Manpower, 'the shortage of skilled manpower will continue unless social stigma attached to blue collar jobs is erased and such jobs command competitive wage and generate dignity in labour' (Malaysia, 1981b: 56). Some have argued that a marked preference for white-collar jobs is associated with current attitudes towards manual work, and that such attitudes are substantially influenced by existing social stratification. But, more significantly, the latter is associated with one's position in the occupational hierarchy and thus one's income. As indicated in Table 7.10, most of the blue-collar jobs are paid even less than those in the clerical and related occupations category.

According to the 1986 Industrial Surveys, the gap between blue- and white-collar salaries was large: the clerical and related occupations category was paid $914 per worker, while the directly employed skilled worker was paid $650 and the skilled worker employed through labour contractors was paid $748 (Malaysia, 1988a: 111). As such, the stigma attached to blue-collar jobs is difficult to erase under the existing social environment. Furthermore, the curriculum in schools tends to be literary and academic and, as in most developing countries, the tendency of even graduate engineers to expect desk jobs and recoil from the prospect of physical contact with machines aggravates the problem (Myrdal, 1968: 1124–31). It is therefore important that these issues be examined in the context of an all-embracing policy on HRD.

Technology development is a knowledge-intensive activity which ultimately depends on the development of human resources. Although the quality of R & D activities depends on the availability of funding, the

TABLE 7.10

Percentage Distribution According to Occupational Categories and Average Earnings in Manufacturing Industries, 1975–1979

| | *Percentage Distribution* | | | | *Average Monthly Earnings ($)* | | | |
|---|---|---|---|---|---|---|---|---|
| *Occupational Categories* | *1975* | *1976* | *1978* | *1979* | *1975* | *1976* | *1978* | *1979* |
| Managerial and professional | | | | | | | | |
| Professional | 1.5 | 1.5 | 1.6 | 1.6 | 1,644 | 1,899 | 2,126 | 2,267 |
| Non-professional | 1.6 | 1.5 | 1.6 | 1.9 | 983 | 1,243 | 1,301 | 1,407 |
| Technical and supervisory | 6.0 | 6.5 | 7.0 | 7.2 | 483 | 513 | 581 | 632 |
| Clerical and related occupations | 7.5 | 7.6 | 7.6 | 7.8 | 331 | 354 | 405 | 432 |
| General workers | | | | | | | | |
| Drivers, conductors, and lorry attendants | 3.0 | 2.1 | 2.0 | 1.9 | 237 | 293 | 332 | 369 |
| Others | 4.5 | 4.3 | 4.3 | 4.4 | 190 | 205 | 233 | 256 |
| Directly employed workers | | | | | | | | |
| Skilled | 28.2 | 28.1 | 29.6 | 29.6 | 200 | 291 | 245 | 271 |
| Semi-skilled | 5.4 | 6.2 | 5.8 | 5.3 | 161 | 154 | 207 | 223 |
| Unskilled | 31.8 | 31.5 | 30.5 | 29.3 | 134 | 147 | 165 | 193 |
| Workers employed through labour contractors | | | | | | | | |
| Skilled | 5.1 | 5.0 | 3.7 | 4.7 | 237 | 247 | 365 | 398 |
| Semi-skilled | 1.4 | 1.6 | 2.2 | 1.9 | 216 | 236 | 251 | 279 |
| Unskilled | 4.0 | 4.1 | 4.1 | 4.4 | 152 | 163 | 178 | 199 |
| Total | 100.0 | 100.0 | 100.0 | 100.0 | 238 | 262 | 306 | 343 |

*Sources*: Computed from *Industrial Surveys 1975–6* and *1978–9*, Department of Statistics, Malaysia.

quality and experience of research manpower involved in R & D in both the public and private sectors is more important. In view of the relatively insignificant stock of high-level manpower engaged in research, R & D work directed towards technological change tends to be very limited, and this in turn limits the capacity for innovation and technological change needed to boost future industrial growth.

In such a situation, coupled with the minimal involvement of the private sector in R & D activities, the development of an indigenous technological capability is severely constrained. The private sector, therefore, has to complement the government's manpower programmes and must invest substantially in the training and upgrading of R & D personnel if it is to build up its pool of R & D personnel to meet its research requirements. The private sector, by increasing its technological assessment capability, for instance, can take advantage of the training component in technology transfer packages for the assimilation of technologies.

The expansion of the country's industrial base will depend, to a great extent, on labour's achievements through technical progress. This is essential for it signifies changes in technology itself or improvements in the art of production resulting from a combination of research, invention, development, and innovation (Thirlwall, 1989: 120). Technical progress can either be externally induced or achieved through the process of learning-by-doing which involves the accumulation of experience by industrial labour at all levels in the course of production. This aspect of technical progress must be given more emphasis by domestic industries, not only because it paves the way for an internal diffusion of technical knowledge but it also improves productivity. Industrial labour will thus improve its skills through on-the-job training, work specialization, and work experience, and become more adept at the job in hand.

The process of learning-by-doing invariably takes place in manufacturing industries, although it may not necessarily be found at all levels of the occupational hierarchy. The process may be restricted to those who are considered technically skilled, including the managerial and professional, technical and supervisory, and skilled worker categories. However, as indicated in Table 7.10, on average these groups accounted for less than 45 per cent of the total labour force in manufacturing industries. In 1979, for example, they accounted for 45 per cent, with the managerial and professional category accounting for 3.5 per cent, the technical and supervisory category for 7.2 per cent, and the skilled category for 34.3 per cent. On the other hand, the semi-skilled accounted for 7.2 per cent and the unskilled 33.6 per cent. In the 1980s, however, there have been some marginal changes in these proportions with the managerial and professional as well as the technical and supervisory categories increasing their shares, thus reflecting the increasing skill content of many industry sub-sectors. For instance, the corresponding proportions in 1987 were 4.2 per cent, 8.9 per cent, 28.4 per cent, 13.6 per cent, and 29.9 per cent (Table 7.11).

TABLE 7.11
Percentage Distribution According to Occupational Categories in Manufacturing, 1983–1987

| | *1983* | *1985* | *1986* | *1987* |
|---|---|---|---|---|
| Managerial and professional | | | | |
| Professional | 1.9 | 2.1 | 2.1 | 2.1 |
| Non-professional | 2.2 | 2.4 | 2.2 | 2.1 |
| Technical and supervisory | 8.3 | 8.9 | 9.0 | 8.9 |
| Directly employed workers | | | | |
| Skilled | 25.1 | 25.5 | 25.7 | 25.1 |
| Semi-skilled | 9.4 | 10.3 | 10.8 | 11.4 |
| Unskilled | 27.5 | 25.3 | 25.3 | 26.8 |
| Workers employed through labour contractors | | | | |
| Skilled | 3.7 | 3.3 | 2.7 | 3.3 |
| Semi-skilled | 2.2 | 2.1 | 2.6 | 2.2 |
| Unskilled | 4.0 | 3.7 | 3.9 | 3.1 |

*Sources*: Computed from *Industrial Survey 1983* and *1985–7*, Department of Statistics, Malaysia.
*Note*: The above categorization does not include clerical and related occupations and general workers.

If technical progress is to be enhanced, particularly through learning-by-doing, it is the unskilled and the semi-skilled that need more attention. This is relevant not only in the context of upgrading their skills but also in terms of increasing their occupational mobility and incomes. The fact that substantial income differentials exist particularly between the unskilled on the one hand and the technical and supervisory and the managerial and professional categories on the other (see Table 7.10) is indicative of the need to examine the structure of earnings within the context of HRD. It is crucial that the semi-skilled and unskilled be given more opportunities to upgrade their technical skills so that their incomes too will increase. There must also be opportunities for promotion, security of employment, and a good working environment.

However, in the early 1990s, the recruitment and promotion system within most manufacturing enterprises still lacks a channel or incentive for these categories of workers to pursue a career without recourse to the formal education system. The accepted route in career development is to pursue a university degree rather than undergo short courses to improve one's skills and proficiency on the job and acquire knowledge through non-formal education; both the wage structure and promotion system are based mainly on the attainment of formal education and the opportunities based on non-formal education are almost non-existent.

The weaknesses in the system must be rectified before a pool of highly skilled industrial workers can be built up. The whole process must give meaning to labour's participation in the country's industrialization programme in that a conducive environment is created for the acquisition of the necessary skills and attitudes for a new industrial order. A conducive environment requires sufficient room for occupational mobility for the majority of industrial labour since 'real progress comes only through movement to a new job involving more in the way of skill, responsibility, independence, and income' (Reynolds, 1964: 390). A skill development programme should therefore create opportunities for upward mobility and facilitate the process by adopting appropriate strategies of training, retraining, and skill upgrading, while placing less emphasis on formal credentials. Even in the European Community, what has been termed as 'continuing education' has been given a great deal of emphasis as this type of education is perceived to be crucial to the current and future technology development within the bloc (Tyson, 1986: 78–80).

The upgrading of skills is therefore an important ingredient in achieving a reasonable rate of technical progress. This is particularly important considering that the country's labour force is relatively youthful in character. In 1980, for instance, 13.1 per cent of the labour force was within the 15–19 age-group and 48.1 per cent in the 20–34 age-group. In 1985, the corresponding percentages were 11.1 and 49.4 per cent amounting to 658,000 and 2.9 million people respectively. In 1990, the corresponding percentages were estimated at 9.9 and 49.5 per cent with corresponding annual growth rates of 0.4 and 2.8 per cent during the 1986–90 period (Malaysia, 1986a: 136). Furthermore, a substantial proportion of new entrants into the labour force are ill-equipped in terms of technical skills. One indicator of this is that unemployment is concentrated among the young, especially those with secondary education. In 1987, for instance, 70 per cent of the unemployed were aged below 25 years, while 69 per cent had secondary education. An increasing number of the unemployed had also completed tertiary education, with the share rising from 1 per cent in 1980 to 3.6 per cent in 1987 (Malaysia, 1989: 90).

## Development of Technological Capability

One of the major constraints in domestic industrial expansion is the relatively undeveloped technological infrastructure in the country. This arises partly because of Malaysia's late thrust towards industrialization and partly because of its long-standing reliance on traditional export commodities. Both, unlike in the resource-poor Asian NICs, create a vacuum with respect to indigenous technology development for industry. Domestic industries, even from the import-substitution phase, have to a large extent been dependent on industrial country technologies, such that there has been no compelling need to develop their own technological base or research facilities. This leaves the manufacturing sector as a whole without a strong tradition for R & D activities in either product or process development. Since it is very costly and risky to develop new technologies (as well as new products), one acceptable option for developing countries like Malaysia is to import most of their technology needs from the industrial countries.

This, on the other hand, creates an environment in which developing countries have to compete among themselves to attract foreign investment which is expected to bring in the technology, consequently increasing the costs of acquiring both the capital equipment and technical know-how. The increasing complexities of commercializing technologies within the industrial countries tend to complicate their acquisition by developing countries, not to mention the constraints arising from the deficiency of domestic technological competence in developing modern industries. This acquisition of technology becomes even more complicated given that local manufacturing enterprises are relatively small, particularly those owned and managed by Malaysians.

The presence of large-scale and diversified firms makes it possible to finance the substantial R & D and capital outlays required to compete in new technology markets in the world of imperfect capital markets. Meanwhile, a highly concentrated industrial structure may pre-empt many opportunities for learning and technology diffusion, which could occur if technological capabilities were evenly distributed among a larger number of firms. For example, South Korea has a highly concentrated industrial structure dominated by the large conglomerates (*chaebols*) while the small- and medium-scale enterprises consist largely of suppliers to these conglomerates. In contrast, the small- and medium-scale enterprises in Taiwan are the most dynamic entities. Concentration levels are relatively low and there are only a few large diversified firms in technology-intensive industries.

The experience of Singapore's electronics and computer industry has amply demonstrated the importance of relating technology development to HRD. Since 1979 Singapore has pursued a clear policy objective of economic restructuring that covers all areas of human and material resource technological upgrading. In the above-mentioned industry, most firms use on-the-job training and management training programmes to prepare employees for their jobs, and much has been learned from the multinational firms (Hakam and Chang, 1988: 181–8). In this context, the modernization and rationalization of the manufacturing sector still depend on imported technology. For most industrializing economies, the long-term strategy is effective only when the process of unbundling such technology is thorough, and this requires mechanisms that will absorb and adapt the imported technologies, and where possible, enable innovation, thus leading to a minimum utilization of foreign expertise.

In the short and medium term, technological capability has to be enhanced by placing stronger emphasis on technology acquisition and adaptation by locally owned firms rather than technology generation, although the latter should remain the final objective in the long run. The literature on technology development has also distinguished between 'generative technology transfer' and 'consumptive technology transfer'. The former refers to transfer of technology, the utilization of which satisfies human needs and, more importantly, has the potential for further generation of technology. This type of technology transfer is critical because the use of knowledge and tools for expanding the domestic industrial base to make tools for stated goals is the crux of the technology transfer process. Consumptive technology transfer, on the other hand, refers to transfer of technology which cannot be applied to satisfy present and future human needs without the technology itself being consumed or exhausted, and thus may not have any real potential for generating any further technology (Yankey, 1987: 45). The development of indigenous technological capability must therefore include the ability to select from available technologies, to master imported technology, and to introduce a degree of originality in the production of products or processes as well as to diffuse such technology throughout the entire industrial sector. These are critical areas which have to be clearly identified and acted upon by the domestic manufacturing enterprises.

## Balancing Imported and Indigenous Technologies

A critical issue that must be addressed is the choice between timely access to new technologies and the ability to develop, if appropriate, such technologies indigenously. Importing technologies from the industrial countries would provide readier access to the state of the art, but it could be both costly and risky in terms of perpetuating an external dependence. Furthermore, if MNCs were the suppliers of technology, they would generally assert that technology can be transferred successfully only through a whole technology package covering the many stages of an investment project, including integrating the technology with management, marketing, and financial skills apart from capital goods, industrial property rights, and technical know-how. If the domestic capability to assess technology is lacking, then the whole process of technology transfer is normally dominated by the technology supplier and the only substantial negotiation is that which takes place between the latter and the host government (Omer, 1988: 29–37). Even then, the host government may not possess the essential expertise and technology assessment capability, therefore leaving the technology licensee or buyer at a disadvantage.

While domestic manufacturing enterprises in general still need imported technologies, they need to be able to integrate foreign technology procurement with the strengthening of their own R & D capabilities (Djeflat, 1988: 149–65; Succar, 1987: 375–95). The issue here is, again, related to the preparedness of domestic enterprises. In order to enhance their technology development, these enterprises must possess the capability to minimize the imbalance between importing foreign technologies and developing domestic technological capabilities, and to enhance their complementarities. There are obvious limitations to the extent of technology transfer through either direct foreign investment or technology licensing. Neither approach will allow much autonomy to domestic enterprises in the area of product development or design. However, unbundling technology licensing will permit greater scope for domestic firms to learn the mechanics of product design through reverse engineering. Significant design modifications and especially innovative new designs require qualitative skills and intimate knowledge of consumer or industry requirements in the industrial countries. This is a critical aspect of technological enhancement particularly for developing countries embarking on export-led industrialization.

Any reduction in a country's technological dependence would theoretically require the minimum import of foreign technology, and the maximum utilization of domestic technology. However, in terms of practical policies, this depends to a large extent on indigenous technological capability, and the costs the country is prepared to bear in the short and medium term to achieve the desired objective. It is therefore important to note that

> an indigenous technological capability is a necessary condition for the evaluation of technology to be obtained from abroad, for the effective utilisation of the transferred technology, for its adaptation to local conditions for getting better terms for the transfer in negotiation with foreign enterprises and for the generation of 'appropriate' indigenous technologies. In other words, indigenous technological capability is not an alternative to transfer but a necessary condition for it (United Nations, 1983: 67).

There are bound to be short-run costs in terms of loss of output and incomes, not only because of the industrial adjustments and rationalization that have to take place to achieve the above-mentioned objective but also because of the new investments in HRD. The experiences of many countries aspiring to become industrial nations have shown that such a strategy is more likely to succeed for some countries than others, and small countries, such as Malaysia, with neither the highly skilled manpower, industrial experience nor an engineering-based industry sector are in no position to aim for absolute technological independence, but with appropriate planning and strategies may aim for partial independence. The country may aim for self-sufficiency in some sub-sectors within a particular industry or in other industries where the potential for success is greater.

An important issue related to technology development in most Third World countries is the acquisition of the capability to utilize existing technologies and to increase productivity. Such an industrial strategy implies that domestic manufacturing enterprises must be able to establish better production facilities while utilizing the experiences gained in production to adapt and improve the technologies in use. A basic approach arising from this strategy is to build upon the technologies that can be accessed from abroad, while nurturing domestic capabilities in areas in which domestic industries and expertise have an edge. Access to technology developments is certainly critical, and this would inevitably enhance economies of scale and allow lower break-even points. Related to this, it has also been argued that a competitive edge in automation must be developed within domestic enterprises (Paul Low, 1991: 97–111). Such automation can be accomplished through a number of approaches, including flexible and advanced machining and processing of more sophisticated products; the development of new capabilities through the adoption of advanced production systems among the domestic capital goods manufacturers; and the adoption of total quality control.

A study on Brazilian company executives concerning their perceptions of the advantages and disadvantages of acquiring domestically developed technologies as opposed to imported technologies indicates that the latter were perceived to be superior, but domestic technologies were seen as being better suited to the local market environment. This implies, first, that vendors of imported technologies must take appropriate actions to improve the adaptation of technology to suit local conditions, and secondly, the state should continue to play a major role by training researchers and even supplying the capital equipment necessary to produce competitive technologies (Christensen and da Rocha, 1988: 5–16). The production of local alcohol from sugar-cane in Brazil is a success story culminating in increased domestic production which simultaneously supports the development of vehicle engines using alcohol exclusively as fuel. This programme not only reduces Brazil's dependency on imported oil but also allows the expansion of the sugar-cane industry (Siemsen, 1988: 143–7).

Technological capabilities within an economy cannot be developed spontaneously, and since the accumulation of any one capability takes time and experience, the sequence in which various capabilities are developed is crucial (Chacko, 1986: 245–52). At the same time, the required capabilities change as an industry or economy matures because of changes not only in existing capabilities but also in demand patterns. It is in this sense that selectivity becomes a critical element in the process of technology development. And since markets respond differently in different environments, technology development policies have to be tailored to accommodate national needs and sector-specific circumstances. Whether new capabilities should be promoted in an industry depends not only on existing capabilities within a particular industry or in related industries but also on the relative costs and benefits of using local or foreign inputs, now and in the future.

In this respect, the National Science and Technology Policy states that the country's policy 'shall focus on the promotion of scientific and technological self-reliance in support of economic activities through the upgrading of R & D capabilities by the creation of an environment conducive to scientific creativity and the improvement of scientific, educational and other relevant infrastructures' (Malaysia, 1986b: 1). However, technological self-reliance in this context, unlike self-sufficiency, does not mean the ability to produce all the necessary technologies, but the ability to produce some of the technologies needed in other countries, which could then be exported to finance the import of technologies not locally available. Given the existence of technological interdependence and the importance of autonomous decision-making capacity, technology exchange on the basis of mutuality of interests thus appears to be an important option for self-reliant development.

Thus, according to the United Nations Economic and Social Commission for Asia and the Pacific (UN ESCAP, 1988: 110–11):

> Achieving a balance of trade in terms of technology content appears to be the most desirable strategy for sustainable development. In other words, every country, however under-developed it may be would require the make-some-and-buy-some technology strategy for long-range development. Technological self-reliance can therefore be achieved through the ability to produce some exportable technologies to finance the import of other technologies. A systematic assessment of technological needs for import, consumption and export is one of the prerequisites for using this strategy. The experience of some newly industrialized countries show [*sic*] that a country must first select and buy some mature technologies from other countries and digest them in the socio-economic milieu through an evolutionary learning-absorption process. It means learning to maintain the facility properly, developing capability to replicate it, then adapting it to suit local conditions, subsequently improving it and finally going in for creation of new technology. In order to buy technologies from abroad the country must specialize and acquire the capability to develop and commercialize some technologies of its own which can be sold in the international market.

In the technologically advanced countries, the performance of scientific and research institutions is based on the vast amount of knowledge that has been accumulated and on constant knowledge-generation in their societies. Developing countries, on the other hand, have neither a modern scientific tradition nor the organizational capability to emulate the experiences of the industrial countries. In order to acquire technological self-reliance, they must therefore have the capability to assess the degree of technological dependence measurable in terms of the technological content or technological value-added of the different goods and services being produced. Secondly, they must also have the capability to assess their technological needs and capabilities, thereby identifying the technological gap and the strategies needed to reduce the gap.

These two prerequisites suggest that developing countries must also possess an autonomous decision-making capability in the selection and management of imported technologies. This would be preceded by an understanding of the attributes of technology and their implications so that an industrializing economy can upgrade its ability to master all stages of technology production in areas purposely selected to take advantage of domestic resource endowments and specialization opportunities, thus improving the country's bargaining position in technology purchases. This will subsequently allow such an economy to formulate and draw up a technology plan, involving the integration of such a plan with the national economic plan so that technological considerations remain important in development programmes (Sharif, 1986: 110–11).

At the firm level, achieving technological self-reliance is a long-term dynamic process which requires strategic and organized efforts on the part of the firm to enhance its technological capability. Having achieved a certain degree of technological self-reliance, the firm must then ensure that this capability be sustained in order to remain competitive and keep abreast with developments in new technologies or products. To do this, a substantial budget might be essential to upgrade its technical manpower through the recruitment of expertise and the formulation of appropriate training programmes. However, it has been argued that a critical weakness of local firms is their inadequate attention towards the building up of technological capability. Thus, according to Hamzah Kassim (1988b):

> Not understanding the concept of technological self-reliance they continue to seek assistance from foreign technology suppliers. In effect, they have substituted imported technology expertise at the expense of local technical resources. From the view point of local entrepreneurs, the development of industrial units is just the importation of foreign manufacturing facilities. To promote this type of transaction, they will resist any attempt by the government to intervene in the technology acquisition process.

Nevertheless, a small number of firms which have emphasized self-reliance and devoted their resources to R & D activities and skill upgrading have enhanced their capability not only in determining the choice of technologies, but also in negotiating better terms of transfer arrangements, supervision of plant construction, and sourcing of local expertise and inputs. In the process, these firms find themselves at a higher level of technological capability.

In examining the need for achieving a balance between imported and indigenous technologies, it is also important to note the reasons for direct foreign investment or, more specifically, the reasons the MNCs are investing in Malaysia. It can be deduced that there is no way these MNCs will transfer technology to domestic enterprises unless there is a coherent and consistent policy on technology transfer and the backup of an assessment capability on technology. In this respect, it is appropriate to extract some of the views expressed by the Malaysian International Chamber of Commerce and Industry, which represents foreign enterprises in the country:

> MNCs and DFI are reservoirs of technology. Their expenditures on R & D are often larger than similar allocations of many LDCs' government. If properly harnessed the MNCs can therefore be instrumental in accelerating the technological progress of many LDCs. Unfortunately, there is often a misreading of each other's aspirations and requirements. Host governments feel the urgency to accelerate the transfer of technology. MNCs, on the other hand, jealously guard their technology which determines their competitive edge and their very survival in the market place. Very often this has ended in misunderstandings and distortion of the proper perspective of the technology transfer question (Beaumont, 1989: 10–11).

The appropriate choice between foreign and domestic sources of technology depends on an evaluation of associated benefits and costs, using prices that properly reflect relative scarcities. The best or most appropriate source is that which yields the highest net benefit. But in assessing technology, it is necessary to look beyond static cost comparisons in deciding which technological elements are to be imported and which are to be supplied locally. What is needed is a combination of foreign and domestic technological information in ways that will enable domestic enterprises to progress quickly and efficiently along their optimum technological paths.

An economy's ability to provide the necessary technological components depends on the stage of development of the relevant sector and of closely related sectors. Manufacturing enterprises engaged in well-established activities may often acquire technological components locally—either through their own efforts or from other firms. On the other hand, enterprises engaged in new activities generally have to rely initially on imported technological packages including all of their components. As they develop their technological capability, they can unpackage these imported technologies so that they import only the technological components that cannot be efficiently supplied locally. The South Korean experience in this respect is worth noting. According to Ahn (1991: 93–8):

> Korea has relied on indigenous efforts to gain industrial competence through various forms of learning by doing and has emphasised transactions at arm's length in the use of foreign resources. Transfer of technology from abroad constitutes only the initial stage in acquiring technology mastery. Of far greater importance are local efforts to adopt the technology that is transferred and to apply the mastery in other undertakings, thereby fostering locally-based innovative activity.

Discussion about the acquisition of imported technologies principally focuses on how such technologies are transferred rather than on what technological components are being transferred and why they are being acquired. Direct foreign investment, whether in a wholly owned subsidiary or in a joint-venture with minority or majority local participation, is likely to be the only way to obtain the latest technology information from abroad. Although such an arrangement can result in some transfer of technology to a developing economy, it cannot really ensure a full understanding of the technology elements because of the absence of local expertise on the technology itself or on technology management. The lack of local control in such investments can thus have several deleterious outcomes for local technology development.

Apart from direct foreign investment or joint-ventures, technology licensing can also enable the acquisition of product or process know-how. If effectively managed, it can permit more domestic control over adaptations and modifications, especially after the licence expires. It can also broaden the options for sourcing of technological and other inputs far beyond what is possible under either direct foreign investment or foreign management control. The main constraints with licensing lie in the capability to absorb the foreign technology and to keep abreast with advances in that technology.

Direct purchases of capital goods provide another way of acquiring the means of production without the attendant constraints of licensing and direct foreign investment. If they can be utilized as models for reverse engineering to produce the machines locally, such direct purchases can be an effective means of enhancing domestic technology. However, in general, the new operating environment for capital equipment requires changes that the instructions do not cover because they pertain to the original environment from where it is imported. Purchases of technical assistance can be utilized to upgrade technological information and understanding to complement an economy's capabilities in production, investment, and innovation. The advantage of such purchases is that they may be easier and quicker forms of technology transfer than if domestic firms were to develop such technologies themselves locally, possibly making costly mistakes in the process. But the disadvantage of purchasing technical services is the tendency towards persistent dependence on such purchases and the consequent neglect of building up domestic capabilities in providing those services easily, and in ways that are more appropriate for local requirements.

### The Ability to Acquire, Adapt, and Innovate

Given that manufacturing industries are expected to provide more highly skilled employment opportunities in the 1990s, economic planners therefore have the important task of formulating strategies to develop new technical skills and upgrade existing ones in the labour force. This is all the more significant since the import-substitution phase of Malaysia's industrialization programme based on consumer goods industries has almost come to a close. There is not much room either for further expansion of export-oriented industrialization if it is mainly concentrated on resource-based and labour-intensive industries.

Moving into the next phase of the industrialization stage, that is growth based on high-technology, knowledge-intensive industries, not only involves the capacity of the economy to accumulate capital and utilize it effectively, but also the capacity to develop new technical skills, apart from continuously upgrading existing skills. New skills, which are important determinants of labour's ability to acquire and adapt imported technologies, must be acquired to cater for the growing needs of these industries and to sustain the momentum of industrial growth if more employment is to be created. The acquisition of new skills must also encompass the capacity to innovate new techniques of production and to introduce new and appropriate products. For instance, in the automotive components industry, new materials such as advanced ceramics, special alloys, polymers, and composite materials are increasingly utilized. As the proportion of high-technology components used increases and their production costs become relatively lower, domestic component manufacturers will have little choice but to keep abreast with these new products and processes or face the possibility of losing their markets to imported components (Ahmad Tajuddin Ali, 1989).

At the same time, the dependence upon imported technologies from the industrial countries needs to be reduced. This issue needs to be emphasized because of the imperfect nature of the international market for industrial technology where 'the cost of acquiring technology are [*sic*] frequently bloated by the manipulation of prices for transactions between constituent units of transnational companies; technology contract clauses that restrict the buyer's exports and require purchases of imported inputs from the supplier' (World Bank, 1979: 65). Manpower planning must take cognizance of this issue, and therefore such planning should become a coherent component of the overall development programme. Although it is only part of the overall development effort, its principal purpose of providing the trained human resources necessary to achieve national development objectives is very important.

Any programme to upgrade technical capability has to consider the adaptive skills of the existing industrial labour force particularly in an industrializing country such as Malaysia where in the early 1990s only about 20 per cent of the total labour force is engaged in manufacturing. The absorption of labour into manufacturing industries, which generally require personnel with an engineering or technical background, cannot be completed smoothly because of the adjustments needed in a totally new environment. Such adjustments take time as they also require on-the-job training, particularly for those who are newly recruited into the industry.

Furthermore, most of the capital equipment or machinery that is imported from the industrial countries is so modern and sophisticated that a local operator requires a longer period of training compared to his counterpart in the industrial countries. In the industrial countries, technological innovations which are widespread in most industries are often directly related to the kind of techniques which have preceded them; in some industries, these innovations originate from the shop floor. This suggests that the need for adaptation by the labour force is less abrupt than in Malaysia. It is also for this reason that in many large manufacturing enterprises, the demand for the technically skilled (including the top management) has to be filled by expatriates.

The experiences of Japan and the Asian NICs suggest that a controlled policy towards technology imports can succeed in securing many of their advantages while avoiding the worst consequences. To be successful, a technology policy must succeed in breaking the ties with industrial country companies, tastes, and products which technological dependence has built up. The more integrated the ties and interests that have developed in the system, the harder it is to make this break. While it is easier for developing countries such as Malaysia to break the ties at an earlier stage of their industrialization, it is more difficult in another sense: they lack the technological capability to make the break. Technology assimilation in South Korea, for instance, has been achieved through a succession of technological efforts over time, largely undertaken by domestic firms to extend their technological mastery and to accomplish minor technological changes. These efforts have resulted in continued and significant increases in the productivity of resources employed in South Korea's industrial sector.

The introduction of technological change must be viewed as a process which has to be guided by a medium- to long-term plan. Macroeconomic policies in general and fiscal policies in particular need to take into account not only the changes at the margin but also at the base. Technological change is most beneficially and smoothly accommodated and absorbed under conditions of steady economic growth. The economic growth after the 1985–6 recession indicates increasing optimism accompanied by a number of policy adjustments effecting structural changes which technological advances are expected to facilitate and generate (Anuwar Ali, 1988).

The state must take the lead in terms of HRD to enhance technological capability and in terms of its ability to forecast technology development and international market changes, although in the case of the latter, there is much to be done in order to enhance the government machinery. Economic development will always depend on the internal innovative capacities of a society, for it is human resources rather than capital equipment that creates development. The pace and pattern of technological change, therefore, depend critically on the size and expansion of the pool of skilled labour, technicians, engineers, technologists, and R & D personnel. Hence the crucial importance of a sound infrastructure relating to education and technological and scientific research. The rapid and sustained productivity growth in manufacturing that is required to transform the country's industrial base, therefore, depends on the increased utilization of well-trained technical and engineering personnel. The success of the overall growth and development strategy in turn, depends on appropriate HRD. In the early 1990s, educational levels amongst the industrial work-force appear to have improved, although this is entirely at the secondary school level.

The manufacturing sector does not seem to be employing an increasing number of youths from post-secondary or higher education institutions. This appears to be a crucial factor in any future development or educational planning given the priority that planners attach to industrial expansion and the role of higher education. The question, then, is how to absorb more of these graduates into the manufacturing sector and how to rationalize educational priorities so that they match more closely with labour/skill demand in the sector. Policy-makers must take a broader view of the agents of technological change, including policies of training and retraining at all levels, the development of an information structure, the training and exposure of management personnel to the processes of technology development, the allocation of local R & D based on available resources, and appropriate tax incentives. There is also a need for a frame of reference to guide the activities of the national R & D institutions and ensure that these activities respond to the real needs of the national production system.

A set of practical strategies or policies should then be adopted to ensure that the industrial sector has an adequate supply of highly skilled engineers and technicians; this will ultimately influence the extent of the country's technological capability, including innovative capacity and creativity. However, at least in the short and medium term, given the predominance of direct foreign investment in a number of key industries, one has to accept the role of foreign expertise in the country's technology development in terms of both technology transfer and technology absorption.

## Creating a Conducive S & T Environment

Technology development gains momentum when a conducive environment for its popularization is created. The weakness of the domestic technological infrastructure is not only the outcome but also the prime cause of technology dependence, implying that there is no option but to import industrial country technology. Strategies to enhance domestic S & T and measures to improve the terms of technology transfer from licensers, mainly the MNCs of the industrial countries, are likely to be ineffective without examining the problems related to economic dependence. The political–economic structure created by this dependence results in an industrial structure that relies on further imports of industrial country technologies.

There is a vicious cycle in which the weak domestic technological base reinforces dependence, and dependence creates weakness. Attempts to break out of the cycle tend to be thwarted by the attitudes and interests developed as a result of the dependent relationship. Action on S & T alone and on the terms of technology transfer are likely to be ineffective without more general action on economic dependence, because the political economy resulting from this dependence requires further imports of advanced country technologies, capital, and expertise.

In view of the above, it is imperative that a long-term programme to popularize S & T be launched as an integral part of a technology development plan. The programme should aim at a universal desire to instil awareness in every citizen to be creative and innovative at the workplace. This, of course, requires substantial efforts at the planning level and an enormous amount of co-ordination between the public and private sectors. In South Korea, for instance, such a programme has been co-ordinated effectively by the Ministry of Science and Technology, the Korea Science Promotion Foundation, and the Saemaul Technical Services Corps in co-operation with other government agencies.

The programme should ultimately develop a rational and scientific way of thinking within society and discard passive attitudes and practices, so that the importance of S & T in economic development is comprehended by every citizen who can then go on to develop the ability to apply elementary technical knowledge to everyday life. Being an important component for the promotion of S & T culture within the country, such a programme should be effectively integrated into the education system. More significantly, the popularization of S & T must be complemented by programmes at the industry level in order to ultimately accelerate the pace of technology acquisition, adaptation, and innovation within domestic industries. These action-oriented programmes should cover the following areas: the promotion of technology diffusion and application in areas such as the enhancement of innovative capacity among domestic firms; the development of new technology-based firms; the development of the engineering services and consultancy sectors; quality awareness and development; enhancement of product design and development capability; innovation-oriented procurement policies; R & D incentives; and the establishment of competence in strategic and emerging technologies.

However, from the very outset effective recruitment and training policies at the industry level need to be emphasized to ensure the success of subsequent policies towards the acquisition and assimilation of imported technologies. These policies should then become the foundation of any firm's achievements in production efficiency and innovation. Some of the crucial ingredients of such policies should include apprenticeship training for newly recruited personnel, the recruitment of technically qualified production managers, and the creation of an awareness among employees of the need to be innovative. All the above strategies would certainly require a strong lead from the state in terms of both policy initiatives and direct involvement in HRD and R & D activities. Its activities must, at the same time, be fully supported and complemented by the private sector.

# 8
# Policy Implications for Enhancing Technology Development

## Policies to Support Industrial Technology

APART from the creation of new technological knowledge and the acquisition of such knowledge, technology development also entails the enhancement of capacities to assess, choose, and adapt the knowledge created and acquired. When manufacturing firms, especially those locally owned and managed, acquire the ability to master the technologies they utilize, modifications and improvements can therefore be introduced, but these also require substantial R & D activities. If judiciously managed, these activities will generally lead to minor innovations which can ultimately have a greater cumulative effect on productivity than the initial innovation.

However, successful technology development within an economy depends significantly on the desire of domestic enterprises to improve production efficiency and competitiveness in their respective industries. Such an attitude must be effectively nurtured within a conducive environment; this requires policy initiatives that will stimulate the growth of innovative firms. Since technical change is more likely to be adaptive and incremental in nature, particularly in an economy with a small industrial base, small- and medium-scale firms can display just as much technological dynamism as their larger counterparts. As such, the role of small- and medium-scale enterprises must also constitute an important aspect of technology development initiatives.

However, firms are generally adverse to spending on R & D as well as technical and skill training if they are unable to reap immediate benefits for themselves since, for most of them, their time horizon is generally short. It is therefore imperative that the state, through direct or indirect intervention in terms of S & T planning, attempts to deal with such a vital issue. Technology, being a cumulative asset, requires commensurate investments to ensure its continuous generation, productive utilization, maintenance, and replacement. Difficult choices have to be made, at any point of time, between immediately productive investments, and these in the developing economies largely entail important procurements, and technology acquisition from the industrial countries or, alternatively, investing in R & D facilities, design and engineering infrastructure, and HRD programmes.

This chapter will highlight the major policy implications with regard to technology development within the economy, particularly with respect to the expansion of the industrial sector. In general, the aims of policies designed to promote technology development can be grouped into three principal categories:

1. To strengthen the demand side, thereby creating market needs for technology.
2. To enhance the supply side, thereby increasing technological capabilities.
3. To provide effective linkages between the demand and supply sides, so as to induce and ensure that adaptive and innovative activities are successful technically and commercially.

If Malaysia is to promote technological progress effectively, proper co-ordination between supply and demand must be given the utmost consideration. As emphasized in Chapter 6, this is the area in which the state must play an affirmative role. Unless there is a competitive environment in which firms perceive that improvements and innovations of products and processes are critical in order to survive and grow, very few firms will venture into such activities as they are usually risky and require a long gestation period before results are realized. At the same time, technological expertise established in public research institutes and the universities not fully utilized in the absence of industry demands for such services. This, therefore, implies the importance of public sector–industry linkages as well as appropriate responses from both the research institutes and the universities to industry needs.

Therefore, S & T policies should be an integral part of the overall industrial strategy that shapes the industry mix within the economy. The demand and linkages may be brought about in a short time span through industrial policies, but the supply of technological capability, particularly manpower development, can only be built through long-term planning and substantial investments. Rothwell and Zegveld (1981: 1) termed this synergy as the 'innovation policy'; it is essentially a fusion of S & T policy and industrial policy. According to them:

> Science and technology policy has been in existence for many years and has traditionally consisted of the patent system, technical education and the promotion of basic science and applied research within the scientific and technological infrastructure. Public policy for industry is also of long standing, being better expressed in some countries than in others. It consists of such measures as industrial restructuring, tariff policy, tax policy and investment grants. Today it seems obvious that the two should be closely related.

Ultimately, it is the supply of a well-trained labour force, which will identify and absorb imported technologies and bring about domestic technology development, that enables a country to sustain its competitiveness. Although a competitive market structure calls for technological

FIGURE 8.1
Support Policies for S & T Development in Malaysia

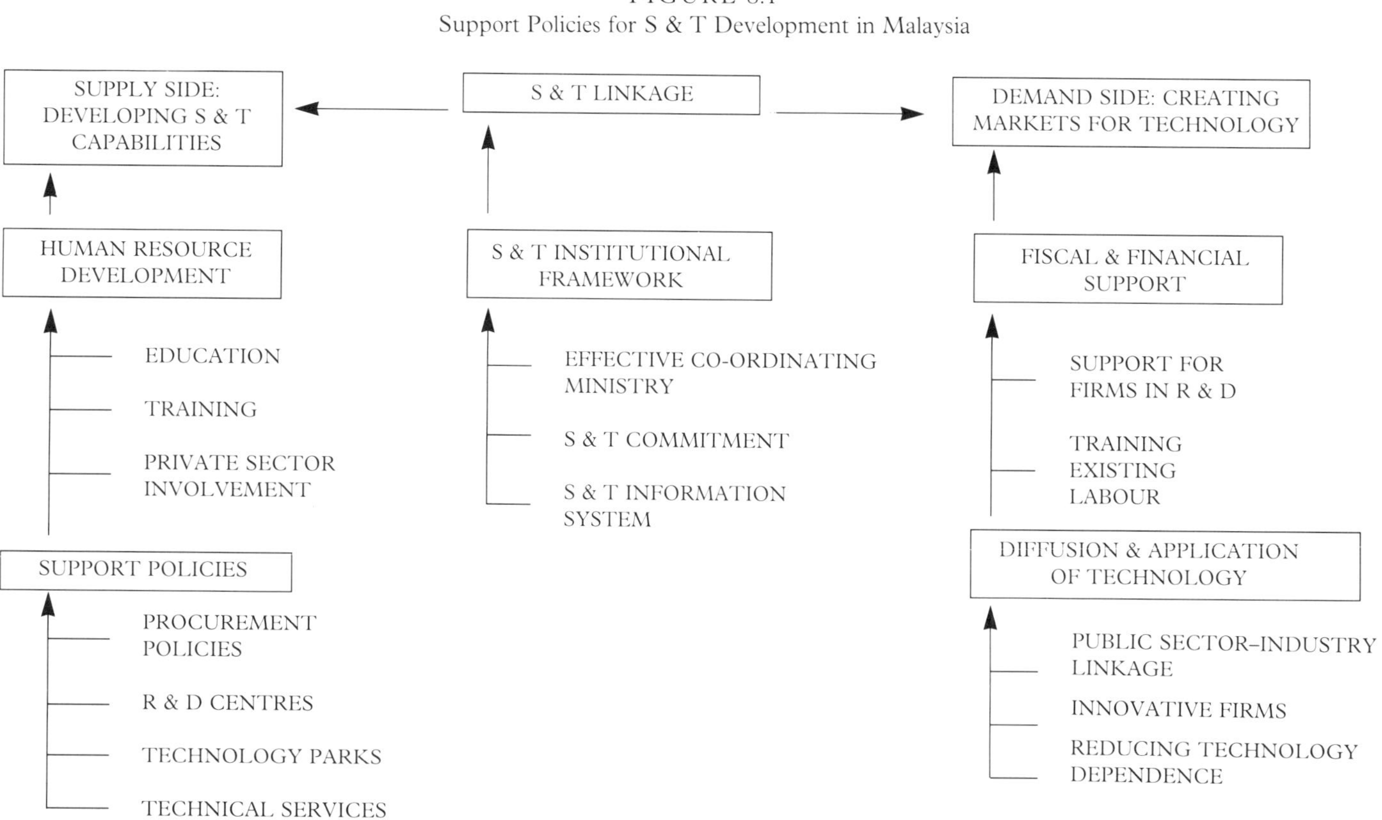

assimilation and innovation, no country can be expected to expand its industrial base without the critical mass of indigenous technological capability, and thus the required and sustained manpower development. Finally, despite the presence of both demand for technological innovation and supply of capabilities, few innovations or even adaptations can be expected to take place unless an effective S & T management system is already in place, linking both the demand and supply sides, that is, a linkage that facilitates the assimilation and innovation processes. It is the absence of this linkage that explains why industrial development in Third World countries often fails to bring about innovations in spite of the strong demand for such technical progress. Although Malaysia may be relatively better off than most other developing countries, there is still a substantial amount of work to be done in these areas.

In the Malaysian context, the major support policies need to cover three principal areas:

1. Policies to develop technological capabilities, that is on the supply side, cover areas in HRD and public sector support policies. Under HRD are strategies related to educational and training institutions and private sector involvement in training. Under public sector support policies are strategies related to procurement policies, industry-specific R & D centres, the establishment of technology parks, and technical services for small- and medium-scale industries.
2. Policies to create markets for technology, that is on the demand side, cover areas in fiscal and financial support and diffusion and application of technology. Under the fiscal and financial support programmes are strategies related to direct support for firms to enhance R & D activities, the training of industrial labour, the creation of special funds for industry to train manpower, and incentives for firms to establish their own R & D. Under the diffusion and application of technology programmes are strategies related to the enhancement of public sector–industry linkage, support for innovative firms, the reduction of technology dependence, and the role of the MNCs.
3. Policies to integrate both the supply and demand sides, that is the linkage provided by the S & T institutional framework, cover areas related to the establishment of an effective co-ordinating ministry, commitment towards S & T, and a S & T information system.

Figure 8.1 summarizes the S & T linkage that is essential to develop indigenous S & T capability and to create the environment necessary for such development.

## Human Resource Development: Education and Training

In the medium and long term, the labour supply constraints, in terms of the shortage of highly skilled and technology-oriented manpower and the mismatch between the supply of and demand for middle-level technical personnel, have to be reviewed so that the role of the various educational and training institutions can be redefined to accommodate the changing industrial needs of the economy. This was appropriately emphasized in the Fifth Malaysia Plan (Malaysia, 1986a: 272) which stated that:

> In the attainment of indigenous competence in S & T, the institutions of higher learning have a major role to play. Such institutions assist in producing middle and high-level manpower capable of absorbing imported technology, executing R & D, generating indigenous technology, and in adapting research results in economically productive ways. Manpower development for the scientific and technical sector is currently faced with the problems of inadequate coordination and linkages between the development planners and the users and producers. In the coming years, there will be a change in emphasis among academic disciplines, and new fields of teaching and research will continue to emerge in response to national needs. It is necessary, therefore to provide the means for the universities to carry out effectively their teaching, research and service responsibilities.

While the local universities are responsible for the upgrading of their research expertise and R & D capability and for the training of high-level engineering graduates and personnel, the polytechnics are responsible for the training of technical assistants and technicians, and the industrial training institutes for the development of skilled workers.

In defining the economic role of the education system in terms of meeting the goals of a rapid industrialization process, it is crucial that a number of issues be taken into consideration. First, it is important that courses offered at the various institutions are relevant and consistent with the expanding needs of industry. In this particular area, some have argued that a major source of skill supply rigidities is actually the public sector itself. For instance, public sector training institutions are:

> ... not as responsive to market changes as they ought to be: curriculum planning takes too long, course contents are often not relevant to industry needs, good trainers are often in shortage due to [a] rigid, unattractive remuneration system compared to the private sector, and quota restrictions may apply, etc. A major policy focus of the government should therefore be how to make our existing system of skill supply more flexible and market responsive (Wong, 1989).

In this regard, the Cabinet Committee on Training (Malaysia, 1991c: 52–3) concludes:

> While the existing skill delivery system, thus far, has been able to meet some of the industrial skill requirements, there is strong indication that it has not been able to meet the manpower demand of industry satisfactorily. According to feedback, public training institutions are not demand-driven, that is, their output [*sic*] have not matched actual industry requirements. It is clear that for the industries to leap forward without experiencing skill bottlenecks, the present training arrangements would have to be reviewed. This is not only to ensure increased skill output, but also to fulfil the qualitative aspects of future manpower requirements. A decision will have to be made to replace the present skill delivery system with one which is more sensitive to market needs.

Secondly, the focus of curriculum development is to ensure that there is a balance between content and skills, that students are trainable for effective participation in the industrialization process. This means that linkages have to be built and strengthened over time between educational planners (including the universities) and industry in terms of manpower training so that both sides can benefit through the exchange of views and experiences. Lastly, a better balance must be achieved between the undergraduate and the postgraduate levels in local universities so that they are sensitive to industry needs; that is, the intake into postgraduate levels must also be commensurate with subjects or research areas that have the potential to be utilized in industry. This is particularly relevant in the engineering and science-based subjects.

Since the education system has its own momentum and is responsible for achieving multifarious objectives, ranging from the economic, social, and cultural to the ethical, any change to be effected, whether drastic or incremental in nature, must be planned from both within and outside the education system. Changes within the education system which are consistent with the socio-economic needs of the country may have substantial repercussions on the capacity of the working population to meet the challenges posed by the next phase of the country's industrialization process. In this respect, a few important pointers must be taken into account.

First, the education system is bound to expand even more rapidly during the 1990s than it did in the previous two decades. Increasing enrolments at all levels will bring about significant structural and infrastructural changes within the education system itself. The present broad-based educational pyramid, with its very narrow apex, should give way to an educational structure that considers the changing needs of the economy, particularly in terms of highly skilled manpower for the modern industrial sector.

Secondly, universities concentrating on excellence in academic and research work should be continuously upgraded, with a shift in emphasis from the arts and social sciences towards the technical and engineering fields. Also required are significant adjustments in the structure of courses offered at the university level. In terms of resource allocation, the expansion of university educational facilities should therefore receive high priority. An important consideration is the enhancement of management education within the local universities as there is also a shortage of professional managers.

Thirdly, the demand for higher education has to be diffused by the provision of a large number of tertiary institutions (technical colleges and polytechnics) which can supply adequate numbers of technical and sub-professional workers, a critical stratum between the secondary schools and the universities. These institutions can also be the avenue for a small proportion to pursue education to the highest levels at the universities. Creating a balance between the demand for the various levels of education must of course be seen in the context of the socio-economic needs of the nation.

Fourthly, changes must also be effected in the reward and incentive structure so that the system will reward those who have acquired the skills needed for the industrialization programme. For example, a gradual change in the incentive structure will act as a catalyst encouraging people to major in studies that are pertinent to industrial production/management systems. An effective measure would be to reassess the wage structure in the public sector that presently rewards academic credentials more than skills such as those in the middle-level technical jobs. Any change in the public sector wage structure is bound to have an impact on industry. This has been recognized by the government in the Sixth Malaysia Plan (Malaysia, 1991b: 178). In 1992, more attractive emoluments and conditions of service and incentives were provided to vocational and technical instructors to attract and retain them in the service.

Fifthly, the inadequate educational facilities to accommodate the increasing demand for places in the universities has necessitated the sending of Malaysian students to overseas universities in large numbers under government and private sponsorship. While this may be unavoidable in the short and medium term, it will be more cost effective and lead to greater self-reliance if industrial manpower is generated by the expansion of local universities.

Indeed, a highly skilled and knowledge-endowed work-force is more critical to national competitiveness in the new knowledge- and information-intensive industries than any other resource endowment. This implies the need for a much higher level of basic education and knowledge than is presently available to meet this challenge. At the same time, there is also the need to refresh and develop, if not to completely renew, skills and knowledge through a process of relearning during a worker's normal working life. It is thus essential that HRD also be applied to the existing industrial labour force not only for the purpose of increasing workers' skills but also to provide opportunities for self-improvement (Anuwar Ali, 1984: 41–59).

To ensure the successful planning and implementation of the above broad guidelines, a manpower forecasting capability has to be established so that the country produces the right quality and quantity of scientists and engineers, technicians and skilled manpower, and that there is a continuous upgrading of technology-oriented personnel. Given the present constraints in both institutional infrastructure and manpower resources, it may be very difficult to forecast future manpower needs with accuracy. This forecasting capability must also be able to provide policy analysis and strategic prescriptions on matters related to human resource development. It is also deemed essential that HRD be effectively integrated with S & T planning. In this respect, the Ministry of Science, Technology and the Environment (MOSTE) need to be actively involved in policy-making for HRD by providing the necessary S & T inputs.

As envisaged by the IMP, private sector participation in policy formulation for HRD will become increasingly vital given its role in the industrialization process. If private sector participation in determining the structure and relevance of training programmes were enhanced, it would help to minimize the problem of matching supply and demand of skilled labour in each industry. Private sector representatives should thus be nominated to the advisory boards of the various training, technical, and engineering institutions, providing the opportunity for more interaction between planners and industry.

This should become an important component of human resource planning for technical skills as public and private sector interaction will first ensure that there is understanding between the training institutions and the industry users of trainees regarding the minimum standards of vocational and technical skills required, and secondly, ensure that there is understanding between the public sector and the private sector on the nature of training to be undertaken by the former while the latter provides training that caters for their own specific needs. Thus, while the public sector must play the leading role, the private sector has to play a complementary role.

At the same time, this kind of private sector participation in policy-making would indirectly strengthen industry associations or encourage the formation of new ones in industries where they do not already exist. With the exception of a few, most of these industry associations are relatively inactive compared to their counterparts in the industrial countries or the Asian NICs, thus weakening their potential in terms of identifying their problems, obtaining and disseminating technological and market information, and forwarding their case to the relevant government agencies such as the Ministry of International Trade and Industry and the Malaysian Industrial Development Authority (MIDA).

## Industrial Training and Support for R & D

Existing skills must be complemented by assimilative, adaptive, and innovative capabilities. In this respect, continuous skill upgrading needs to be given serious consideration if domestic industries are to enhance their competitiveness in both the local and international markets. However, with few exceptions, domestic manufacturing firms generally do not have specific training budgets, although on-the-job training, ranging from a few days to a month, may be provided to new and existing employees. For most small- and medium-scale enterprises, this type of training is almost non-existent. Most training programmes for upgrading skills are provided on an *ad hoc* basis according to the availability of courses and financial allocations (Anuwar Ali, 1991: 59–64).

The above constraint is further complicated by the fact that in certain industries, where specific skills are in short supply, manufacturing firms tend to lose out to their competitors in the face of rapid and continuous labour turnover. In the machinery and engineering sector, for instance, there are many small firms which are under capitalized and offer low-entry pay to their workers. At the same time, the increasing number of MNCs joining the labour market hive off the experienced employees trained in small firms by offering higher salaries. Thus, a situation arises in which small-scale enterprises are effectively bearing the cost of skill training, while the MNCs mainly reap the benefits (Malaysia, 1991c: 30).

Greater emphasis must therefore be placed on retraining of the industrial work-force to provide specialized and up-to-date skills. Retraining is certainly crucial given the rapid changes in technology which require immediate responses at the industry level. Such training can be effectively encouraged by utilizing more of the incentives given for training, for example, the double deduction tax incentive. In this respect, the IMP has proposed that measures be instituted to make it compulsory for manufacturing enterprises to train up to 10 per cent of their staff. Although the implementation of this proposal may be subjected to a number of problems, such training should nevertheless encompass many areas for improving employee numeracy and literacy, imparting skills to unskilled and semi-skilled workers, and advanced skills to skilled workers.

There is already a proposal to initiate a skills development fund under the Action Plan for Industrial Technology Development to finance programmes to upgrade skills amongst workers in the industrial sector. According to the Plan, such a fund will form the basis of a new approach to industrial training; it has been introduced successfully in several countries (Malaysia, 1990b: 50). For example, Singapore initiated its Skills Development Fund in 1979; a levy was imposed on employers which was used for the objective of upgrading skills in manufacturing enterprises as well as acquiring better technology and equipment. The Fund has been successfully managed and it now has a number of schemes to ensure that companies will train their employees effectively; these include the Training Grant Scheme, the Interest Grant for Mechanization Scheme, the Development Consultancy Scheme, and the Initiatives in New Technology (INTECH) Scheme (Chng et al., 1986: 7, 47).

A similar fund to be adopted in Malaysia will be managed jointly by the public and private sectors, and financed through a cess collection from the manufacturing sector amounting to 1 per cent of a firm's payroll (*New Straits Times*, 19 April 1990). However, this cess will not be imposed upon newly established enterprises and small-scale industries as such measures will only jeopardize their potential to establish themselves and to expand their activities. One option is to apply this proposal on a selective basis according to firm size or 'priority industries'. While the fund is aimed at enhancing skills at the firm level, the levy imposed may have a negative impact on the absorption of labour. However, in a situation where wages in industry are increasing, it is best that those who are already employed in manufacturing industries be equipped with new skills.

Besides upgrading existing skills, grants will also be made available from this fund for manufacturing firms to pursue R & D activities considered vital to the needs of the sector. These will enable firms to obtain grants or seed capital to develop new products and processes or to substantially improve upon their existing products or processes. The fund can thus complement the Industrial Technical Assistance Fund (ITAF) announced in the 1990 Budget under which $50 million will be provided by the government to increase R & D activities among the small- and medium-scale industries. The ITAF will be used to reimburse 50 per cent of the total project costs incurred by firms undertaking feasibility studies, quality and productivity improvements, product and design development, and market development (*New Straits Times*, 27 April 1990).

The establishment of the ITAF is particularly meaningful for small- and medium-scale enterprises since they seldom utilize the existing incentives and lack the start-up funds for R & D activities. However, in practice, the constraint is generally related to the issue of accessibility to fund, especially by small-scale enterprises. In this respect, it is also useful to examine the major R & D incentives that have been granted by South Korea and Singapore, and apply them to suit local industries. In order to utilize the grant effectively for as many R & D activities as possible, these activities should preferably be carried out by Malaysian-owned and -controlled firms through their qualified Malaysian personnel, and only when necessary, foreign experts and consultants should be employed to supervise the work. Funding priority should be given to R & D activities which have the potential for commercial application.

One strategy with regard to this Fund is to provide it on a 50 : 50 basis so that half the costs will be borne by the firms undertaking the R & D activities. However, if it is carried out in collaboration with the local universities and public sector research institutes such as MARDI, MIMOS, SIRIM, RRIM, or PORIM, it would be possible to give a 100 per cent grant with a view to increasing the public sector–industry network.

While the state can allocate funds to finance R & D activities in the public sector, other forms of direct support to stimulate private sector R & D activities must also be seriously considered. It is difficult for most of the locally owned firms to initiate or increase their involvement in R & D without the explicit guidance and support of the state. Stronger and more definitive support has to be given so that local manufacturers can launch their R & D in-house since their participation in these activities is still in its infancy.

The principal objective should therefore be to encourage domestic firms, particularly the larger ones, to establish their own R & D facilities which may initially concentrate on modifications and adaptations of imported technologies and eventually, when a critical mass is achieved, enable innovations through their own in-house skills. Innovative activities nevertheless require not only highly specialized and skilled manpower but also fully equipped laboratories, equipment, and the building of prototypes. All these manpower and infrastructure requirements need substantial financial resources and governmental fiscal support. The experience of South Korea provides a useful lesson in this respect. For example, South Korean venture-capital firms have generally assisted small-scale enterprises in adopting appropriate imported technologies rather than investing in riskier ventures. Firms are generally financed by debt rather than equity, and little pressure is exerted on new firms to go public. However, new laws have been enacted to encourage new venture-capital companies to go public and to give them more access to funds (Clifford, 1988: 52).

In order to promote R & D activities, a number of possibilities can be considered to make such activities more attractive to existing as well as newly established firms. For instance, the definition of R & D for tax purposes could be expanded to include engineering and design activities in order to improve the technological absorptive capacity in the manufacturing sector, while expenditure on scientific equipment used exclusively in R & D activities could be made duty-free and tax-exempted. At the same time, the practice of tendering out research contracts by the government or industry to research firms or even research institutes and the universities to undertake research for specific projects should be instituted. This would represent a business opportunity as well as an investment in R & D for the successful firm and would contribute to the long-term development and upgrading of the technological capacity of the country. Such an activity would simultaneously enhance public sector–industry linkages. It may also be a good idea to restructure R & D tax incentives to allow R & D write-offs to be carried forward or backward for several years for ease of financial planning so as to accommodate the long gestation period of research activities before significant results are achieved. Such a measure diffuses the risks taken by firms since the larger the amount of investment outlays required in innovative activities, the higher the risks.

## Promoting Technology Diffusion and Application

An important prerequisite for promoting technology diffusion and application within domestic industries is the creation of networking or collaborative linkages between the public sector and industry. This has been the experience of all industrial countries as well as the Asian NICs in achieving their technological advancement. Public sector research institutes and the universities, for instance, can draw upon their existing expertise and skilled manpower to secure R & D jobs or consultancy work from industry. Each research institute and university will have to upgrade its expertise in its areas of specialization or in new priority areas that are designated by the government. This kind of public sector–industry interaction will create a competitive environment between these institutes and universities so that they will always strive to be in the forefront of R & D in their selected areas of specialization. The difficulty is to translate this into reality.

One must note that technological enhancement is more than simply the exchange of technical publications. It also includes various contact mechanisms and communication activities that are essentially person-to-person linkages. Cutler (1989: 17–24), for instance, concludes that personal communication and technical collaboration are the key factors in the rapid diffusion of high-technology research results in Japan and the United States (Martin, 1988: 211–19; Kuhlman, 1986: 15–19). In this context, it is important that industry responds positively by allowing channels of networking with local universities and research institutes.

During the 1990s and beyond, the universities in Malaysia must play a more positive role in the country's technology development not only because of rapid global technological changes but also because of the country's objective to increase the share of manufacturing in its total output. However, these challenges must above all be met by a clearly defined strategic response from the many relevant sectors, including each of the local universities (Anuwar Ali and Zawawi Ismail, 1990).

This strategic response from the universities will only materialize if certain prerequisites are met. First, the message on the need for a strategic response to rapid technological changes must be heard and clearly understood by the university leadership and its executives, because there must be total commitment on the part of the university towards industrial and S & T development within the country. This leads to selection on the basis of priorities not only in terms of the appropriate intake of students according to industry needs but also increased university–research institute–industry collaboration. Secondly, an environment within the universities conducive for innovative ideas and skills to flourish must be nurtured to ensure that academic talents are duly recognized and rewarded. An innovative culture has to be cultivated if a strong technological capability is to be established within the universities, and in this process, the universities will be recognized as centres of excellence.

Within the existing framework, however, local university expertise generally lacks the vital exposure that is essential for undertaking consultancy services with manufacturing firms. The opportunities that are available are very limited; they are mainly obtained through individual contacts. There is also a lack of demand from domestic industries for the expertise available in the universities or public sector research institutes. This arises mainly because of the dearth of R & D activities within most domestic firms. In the case of small-scale enterprises, and even medium-scale ones, R & D activities are almost non-existent. If there is any such activity, it is principally done in-house without the need to utilize high-calibre expertise. There is also a perception within industry that expertise from local research institutions lacks the technology inputs to cater to industry needs. However, if effectively utilized, university expertise constitutes a major source of technological innovation and technology information.

In order to encourage the development and utilization of indigenous R & D, the rules governing the involvement of university personnel in industrial consultancy work should be relaxed. These personnel should be accorded recognition and incentives when they participate actively in R & D activities. There is also the need for a better mechanism for information flows between industry and the universities/research institutes. While university–industry linkage should be encouraged, such research collaborations should only be undertaken when they are consistent with the principal goals and functions of the university. The latter must be able to preserve a balance between its teaching, research, and social objectives without inhibiting intra-university exchange of personnel and information. The university must therefore ensure that any industry-sponsored research is conducted in an unbiased manner, contributes to basic knowledge, and provides students with worthwhile research opportunities (Brown, 1985: 7–17).

Recognizing the need to maintain a working relationship between the research institutes and the final users of technology, the former should venture out to secure jobs and contracts from industry. For instance, an expert or team of experts in a certain field from within these institutes can be assigned to undertake a certain R & D project for the benefit of a firm or a group of firms within a specific industry. The industry then subsidizes a portion of the total cost of that project and in so doing determines the pace and direction of the project. Instead of funding the entire operation, the state need only subsidize a portion of the project. The proposed public and private sector subsidy proportions could be about 70 : 30 and gradually progress towards a greater amount of private sector funding depending on the number of jobs and contracts secured from the public sector.

In the long term, it is also critical that industry increase its R & D expenditure, although the present low level of private sector R & D capability dictates that the targeted increase be phased realistically over the years. Since investments in R & D facilities and expertise are long-term and involve high risks, this target should then be prioritized according to the demands of each industrial sub-sector. The rationalization of private sector R & D activities must also be complemented by joint efforts with public sector research institutes. At the same time, there needs to be support from these institutes for industry. Such support can take the form of assistance in prototype development, provision of incubator centres, establishment of good manufacturing practice, and improvement of the information dissemination network.

In initiating any extensive industrial R & D programme, one must consider the rapidly accelerating costs of all types of research activities. There should be an awareness of the advantages of concentrating resources on teams that are achieving excellence, and the need, in view of the accelerating volume and pace of change, for being systematic about strategic research and development. Selectivity in R & D has also been motivated in many instances by a desire to reduce public expenditure as an overall national policy; this has forced choices on distribution of funds and selection of research teams. This trend has been the norm in major strategic programmes in Japan, the United States, and the European Community, especially in micro-electronics and information technology. It has also been the strategy adopted by individual research institutes in both the public and private sectors where there has been a recognition of the advantage of building up research strengths.

Malaysia cannot assume that the industrial countries are willing to sell their newer or frontier technologies to locally owned manufacturing enterprises for they naturally fear losing their competitive edge in the international markets. Therefore, if domestic industries are to reduce the technological gap, they need to improve their technological competence and be able to innovate as fast as possible. Given that the innovation process is very complex, since a large number of inputs are required, many factors must be considered in order to ensure the effectiveness of such innovation. First, as some have argued, at the industry or firm level there should be a product champion who is committed to the innovation project throughout the technology development stage. Secondly, effective interaction must be developed by the product champion with key decision-makers in government agencies. Thirdly, there must be a long, close relationship within the design team composed of committed experts. And lastly, there must be an organic connection between the technology development and production stages provided by transferring the development team to the manufacturing enterprise (Chaudhuri, 1986: 89–103).

At present, the manufacturing activities of locally owned firms, being generally dependent on imported technologies through licensing or joint-venture agreements, are mainly concentrated in low to medium value-added operations, and there is very few internal innovative undertaking. Moreover, they contribute only marginally to the development of the overall technological capability. This in itself creates a vicious circle in which domestic industries are still heavily dependent on imported technologies and will continue to be so throughout the 1990s. Indeed, there is no denying that, given the present status of the domestic capability for innovation and technological change, the country's R & D programmes to effect the development of indigenous technology will have to be built upon foreign technological inputs, generally through the MNCs.

In the long run, domestic industries have to overcome the technology dependency syndrome by developing their capabilities to select, acquire, assimilate, and adapt imported technologies and thereafter innovate. Although investments in technological innovation are generally risky, long-term, and costly, the average returns on such investments are relatively substantial for innovative firms, and when taken together, even larger for the economy as a whole. Studies on the costs and returns involved in the development of new or improved products and processes in several manufacturing sectors in the industrial countries show average earnings of 25–35 per cent. These high returns on investment are generally obtained through incremental advances and technological improvements of products and processes rather than from major technological advances or radically new products which can earn huge benefits for innovative firms (OECD, 1985: 55).

The state machinery must be deployed towards a long-term objective of technological enhancement by identifying manufacturing enterprises that employ active technology strategies through the utilization of highly motivated and committed managers and producers. The challenge to the state is to stimulate the efforts of such innovative enterprises by providing the right mix of incentives and regulations that will encourage them to expand and flourish. These enterprises may not necessarily be large ones, but could also be small- and medium-scale enterprises. A study by Girvan and Marcelle (1990: 91–107) in Jamaica concludes that small enterprises are as 'intrinsically capable as any other type of enterprise to engage' in active strategies for technology acquisition and learning. This can be effectively done by astute use of relationships with raw material suppliers and other business contacts, backed by systematic in-plant learning and experimentation, and a far-sighted policy of investment in human resources.

To support innovative firms, a programme for product development within the domestic industries can be initiated to encourage them to engage in innovation with the provision of seed capital financing, particularly at the initial stages of such activities. This support programme should be complemented by measures to encourage a greater sense of urgency regarding the quality of products manufactured within locally owned firms.

Apart from innovative firms, the growth of new technology-based firms (NTBFs) must also be encouraged because these are the firms which promote and commercialize ideas arising from technological innovations wrought in collaboration with local technologists or innovators. Firms which give priority to R & D activities and utilize completely new technological expertise must be fully nurtured to grow and exploit their full potential. Particular emphasis could be given to areas such as robotics, electronics, biotechnology, and information technology-related products in telecommunications and office automation. This is a difficult task for developing countries like Malaysia, as it involves the nurturing of a new breed of S & T oriented entrepreneurs whose existence is a prerequisite for the development potential of new firms. Apart from nurturing these entrepreneurs, who can be found in the small- and medium-scale enterprises, this strategy would also require a certain degree of mobility of individuals both in industry and in the universities.

Policy instruments for this purpose cover a wide spectrum of measures: government procurement, subsidies, tax incentives and credits, financial aid for R & D and innovation, supportive scientific and technological infrastructure, technical services, educational programmes, and assistance in obtaining venture capital. In the industrial countries, the growing support for diffusing new technologies to small innovative firms can be seen as an effort to sustain and promote the growth of existing companies, in contrast to creating new firms. Several of these countries now provide assistance and incentives for such companies to upgrade their technology base. This is being approached through cost-sharing schemes in some countries and in others through the creation of technical centres which provide various kinds of service (OECD, 1985: 82).

An important strategy for encouraging and sustaining the growth of innovative firms is the expansion of domestic capabilities in the machinery and engineering sectors as well as engineering consultancy. The local machinery and engineering industry is relatively underdeveloped, and one of the major reasons for this underdevelopment is the lack of design and precise engineering capabilities within the industry as well as in engineering consultancy firms. But there are some capabilities being developed in areas such as fabricating, welding, forging, casting, and machining. It is important that this sector be further improved; and in deepening the country's industrial base, it is essential that local technologists/engineering consultants be able to develop design and precise engineering capabilities for making machines or capital equipment.

There is thus a need to upgrade this sector to the highest level of technology utilization so as to forge a strong linkage with the local foundries and manufacturers. A number of measures can be adopted to enhance development and design capability, including programmes on advanced methods in design development by the use of computer-aided design (CAD) systems and the nurturing of specialist designers in various industries.

Design and engineering consultancy services constitute an important link in the chain of technology development. These services may account for as much as 15–20 per cent of the total project cost, depending on the size and technological complexity of the project. In view of their sizeable financial resources, major public enterprises, such as HICOM and PETRONAS, are in a position to strengthen their internal departments which are involved in detailed engineering, equipment identification and procurement, and construction supervision services so that it becomes possible for them to take on these responsibilities in respect of projects increasingly undertaken within the country. As industrial development gathers momentum and technological capability attains a higher level of competence, the volume of business in engineering and consultancy services may justify these departments being separated from their parent companies and set up as separate entities.

As adequate numbers of experienced engineers, technologists, specialists, and draughtsmen may not be readily available within the country to embark on this relatively sophisticated field of technological activity, establishment of joint-ventures in association with foreign design engineering and consultancy firms from abroad may be permitted, provided that there is increased participation of local people progressively over time. This would require an effective monitoring system.

The challenge facing domestic industries is the need to develop and rapidly expand the country's technological capability. This challenge can only be effectively met if public sector research institutes and manufacturing firms have the capacity to absorb and adapt technologies quickly and efficiently. Technologies for the development of industries are heavily dependent upon mechanical, electrical, and metallurgical engineering considerations. In these areas, since relevant skills and design capabilities for replicating and eventually improving upon foreign technologies are seriously lacking in the country, effective R & D programmes in these areas can therefore make important contributions to the development of the nation's indigenous technological capability. The planning of such programmes must therefore be selective and the choice of industrial technologies and research projects must be prioritized according to current and future needs.

In this respect, the Malaysian government has identified several new high-technology areas that are to be given priority for R & D support, and these include micro-electronics, laser technology and electro-optics, biotechnology, materials technology, manufacturing technology, and software technology. Some strategic activities are also identified, and these include remote sensing, oceanography, applied climatology, computer-aided design (CAD), computer-aided manufacturing (CAM), and computer-aided engineering (Malaysia, 1986a: 273).

In determining priority areas, it may be useful to look at the future development strategy of South Korea as an example. South Korea is seeking to establish an industrial structure which emphasizes the manufacture of products with a high value-added commensurate with its high-quality human resources combined with an optimum utilization of its natural resources and energy. This means that it aims to move towards a 'small but advanced' type of development (as exemplified in countries such as Switzerland, Belgium, the Netherlands, Denmark, and Sweden) requiring the fostering of strategically specialized industries, optimizing the social and industrial system, and promoting the quest for a high-technology society (Choi, 1988).

If GDP growth per annum is projected at an average rate of 7.5 per cent during the Sixth Malaysia Plan (1991–5), annual investments will have to increase at the rate of 7.0 per cent per annum, from $25.9 billion in 1990 to $36.2 billion in 1995 (Malaysia, 1991b: 18). From the experience of the Fifth Malaysia Plan (1986–90), such a vast amount of investment cannot be easily met from domestic capital alone despite the extremely high savings rate of about 31 per cent of GNP during the Fifth Plan (Jaafar Ahmad, 1989). The flow of foreign investment is deemed crucial in achieving this objective. As indicated in the earlier chapters, direct foreign investment has been an important source of capital, management know-how, market accessibility, and technology. Policies on direct foreign investment are therefore important in influencing the nature of technology utilized, and the pace and direction of technological change in the economy. It would seem that technology imports to sustain industrial development, and implicitly domestic technological development, are now widely accepted as a matter of policy. The difference has to do with the extent or degree of reliance on imported technologies instead of indigenous technologies.

The policy mix to achieve an acceptable level of dependency on imported technology depends on, among other factors, the type of technology to be implemented or the speed required to implement the technology, the relative costs of technology imports and domestic R & D capabilities, the relative efficiency and suitability of foreign and local technologies, and the availability of local technical skills. The experiences of Japan and South Korea with regard to direct foreign investment and their advancement in S & T capabilities provide some useful pointers with respect to the policy options that could be emulated by others. Apart from their emphasis on a strategy based on borrowed technology, considerable efforts have been concentrated on the development of domestic R & D as a means of technology absorption and adaptation.

In line with the country's objective of building and developing its capacity for innovative technology, the private sector, while continuing to import technologies through various mechanisms, must endeavour to ensure that technological information and know-how from technology suppliers is actually being imparted. At the same time, it is important that industry assists policy-makers in ensuring that there is effective technology transfer by advising the latter on the requirements of domestic technology licensees in terms of technical assistance fees, training for local employees, purchases of materials and capital equipment, and export conditions. Given that MNCs negotiate with policy-makers, it is therefore important that there is constant interaction between them and local managers or leaders of industry.

Policy intervention is also deemed necessary to ensure that genuine technology transfer occurs, for instance, through the establishment of in-house R & D facilities. Such intervention may take the form of inducement and compulsion; the former refers to fiscal incentives made available while the latter refers to various conditions that MNCs should comply with in areas such as the transfer of basic product or process technologies and the employment and training of local engineers and technicians. It must also be emphasized that technology transfer requires training and skill-upgrading to be performed by the MNCs at the local level, and such training must respond to the needs of the local employees in such a way that their courses are designed or modified to accommodate the local environment (Copeland, 1986: 107–18). In this respect, it is important that the government reviews the policy on joint-venture projects with MNCs so as to assess the benefits accruing from these projects. Khor (1983: 242), for instance, pointed out that

> the foreign partner inevitably enjoys overwhelming control over the technological and financial aspects of the joint venture. The joint venture thus becomes a new institution through which surplus can be channelled out to the multinational corporations via multiple channels such as royalty payments, transfer pricing practices, charges to the head office, 'technological lock-ins' and so on.

What happens then is that the foreign firms relinquish formal majority ownership of assets in the economy, but maintain dominant control over surplus through the use of their technology and through accounting procedures.

Although a few of the MNCs undertake R & D locally and some have transferred technology to domestic firms through subcontracting and supplier relationships, these efforts are in the very early stages. In the main, MNCs remain offshore operations contributing little to local technology development. However, industrial skills required to introduce and operate new technologies can be instituted through vendor training or subcontracting training. In the former case, overseas vendors of machinery or equipment train local personnel to operate and maintain the equipment, while in the latter case, it occurs when a MNC wishes to source its inputs locally and thereby trains local suppliers in quality control and other process-related methods.

## Public Sector Support Policies

Domestic technological capability may also be enhanced through public sector procurement policies which can be utilized to maximize innovative activities within the supplier industries. Such policies must be able to promote as much import-substitution as possible and to enhance industrial capability to produce goods for the public sector. While enhancing this capability, domestic industries will indirectly become more competent in terms of their export capability. The specifications demanded of the domestic industries will greatly influence the type and level of innovative activities that the government wishes to pursue. One specific area of relevance could be related to the Memorandum of Understanding between Malaysia and the United Kingdom signed in 1988 in which the latter agreed to supply the former with a package of military hardware, training, and technology transfer. An important aspect of the memorandum is the provision for the joint manufacture of defence equipment in Malaysia for export to third countries (Seaward, 1988: 85–6).

Utilizing the public sector's purchasing power in promoting domestic industries is not entirely new, and the policy has in fact been adopted since the early 1980s. However, this policy should be vigorously pursued with more specific guidelines to take full advantage of the potential in terms of technology acquisition and innovation. Rothwell and Zegveld (1981: 24), for instance, observed that in the industrial countries, the

government itself is heavily involved in setting the rules of the marketplace in general and is for many market segments a major customer. Government influences demand through regulations and procurement. In fact, a variety of studies that stress the importance of 'demand pull' as the critical input in the process of innovation support this standpoint. If innovation policy is to become more than the traditional science and technology policy, governments cannot neglect the manner in which they influence demand.

In Malaysia, procurement policies generally seem to be devoid of any technology enhancement content. There is also no explicit policy to encourage domestic manufacturers to be innovative. This constraint must therefore be minimized by reviewing existing policies. It must also be consistent with the need to enhance engineering design and consultancy capabilities.

The gathering of a large number of firms, especially within a specific industry, creates a critical mass that enables the pooling of resources to achieve economies of scale in research activities, hence allowing for the optimum utilization of common R & D facilities. The rationale for such an industry-specific R & D centre is therefore similar to the justification for the establishment of industrial centres such as the Iron Foundry and Engineering Complex and the Furniture Complex; the latter is scheduled to be completed in 1992. The creation of such centres requires a lot of effort from both the public sector and industry given that there is always an inherent conflict of interest among firms within a particular industry. This strategy is unlikely to materialize in this country given the current stage of industrialization, but in the long run, the strategy can be pursued.

In Canada, for example, an important strategy that has been vigorously pursued for enhancement of technology development has been the formation of research consortia. These permit the sharing of R & D costs within each industry, a strategy which complements the programmes initiated through university–industry collaboration (Powell, 1989: 1–4). The state's primary role in complementing R & D of this nature can be carried out through a number of other strategies, including the financing of capital equipment or machinery and up to 50 per cent of the operational costs of these R & D centres. The actual R & D activities will be determined and subsidized through industry contribution.

One of the proposed incentives under the IMP is to grant licences for expansion capacity to firms wanting to commercialize their R & D results. Commercialization of R & D, however, can be effectively realized if the scientific discoveries are translated into marketable products and services and there are facilities to provide additional support for scientific research. An important conduit for promoting this type of activity is the technology park or innovation centre. A technology park is an industrial site which ideally promotes interaction and collaborative efforts between universities or research institutes and private firms for the purpose of nurturing the growth of the latter through production improvements and R & D efforts. By providing such a linkage, a technology park encourages the formation and growth of S & T based industries and becomes the channel for the transfer of technology and skills from these institutions to the industries. The establishment of technology parks or innovation centres can be encouraged in suitable industrial areas where there already exists a critical mass of manufacturing activities.

In the United States, for example, high-technology companies evaluate potential relocation sites based on the degree to which industry, universities, and government in these areas co-operate in research and technology development programmes. A co-operative research environment is desired because it makes research findings more widely available and it increases the speed with which application of the research reaches end-user markets (Wigand and Frankwick, 1989: 63–76). A similar recognition of needs can be traced within the European Community which emphasizes the importance of the above-mentioned interaction and linkages (Narjes, 1989: 241–8).

Another advantage of such research-oriented concentration is the possible linkage with financial institutions. In India, for instance, the Technology Development and Information Company of India Limited (TDICI) was established in 1988 to undertake venture-capital activities. More significantly, it was sited in Bangalore, which has become the S & T centre of India with the electronics industry being the main focus of activities. The TDICI, which provides financial assistance to firms involved in the development and/or commercialization of innovative technologies, becomes a partner in such activities, sharing the risks and rewards with the promoters of the projects, while providing active support in terms of project management, either directly or through consultants. Up to the end of 1989, it had financed about forty small- and medium-scale firms in emerging technology areas such as computer hardware and software, telecommunications, biotechnology, pharmaceuticals, chemicals, and materials (Nadkarni, 1989).

Given the importance of small- and medium-scale industries (SMIs), more emphasis must be given in Malaysia to the expansion of R & D facilities and technical and consultancy or extension services to these industries, especially with regard to purchases of the appropriate machinery and equipment (Rahim Bidin et al., 1984: 105). These services could cover areas such as machinery and equipment appraisal, manufacturing process appraisal, cost analysis of technological components, and the sourcing of technology. Of course, these consultancy services must be supported with technological or industrial information encompassing technical data on machinery and equipment, productivity, product specifications, and technology and input sources, thus enabling the SMIs to play an increased role as suppliers of inputs to their larger counterparts.

SIRIM has already established the infrastructure and it could enhance this by establishing a more integrated R & D centre. While it is most desirable to extend these services to the SMIs, much of the failure in doing so is due to the lack of funding and expertise. This is further complicated by the fact that these industries are spread all over the country and are not homogeneous, and thus require expertise in almost all subsectors. An equally relevant factor is the identification of industries in which these small- and medium-scale enterprises, especially those owned and managed by local interests, can be involved. MIDA has identified four such industries and products, including the moulds and dies industry, packaging products, plastic components, and metal stamped parts (Low Peng Lum, 1990).

In the case of the food-processing industry, for example, MARDI's role could also be expanded. The small- and medium-scale processing firms are generally associated with traditional methods of processing which need upgrading to ensure acceptable product quality. In this respect, the Food Technology Division of MARDI could be strengthened in areas such as technical services, project planning, testing and quality control, and technical information.

With their experience and given the necessary funds and expertise, existing agencies should be able to provide better advisory and training services to all industries, especially to the small manufacturing enterprises. These services include keeping up-to-date information on technical know-how, engineering products, machinery, and services available in the country and ensuring its widest dissemination. An important aspect of extension services for the small-scale industries would be to enhance the technical and quality control support measures to meet the quality standards of the users, particularly for those enterprises producing intermediate inputs or parts and components.

One possible mechanism which could be enhanced is the development of a tripolar arrangement, involving the public sector research institutes, the SMIs, and the banking sector. The banking sector is indeed crucial in terms of the financial services needed by industries. In South Korea, for instance, the Industrial Bank of Korea has been actively involved in linking the SMIs with research institutes such as the Korea Institute of Machinery and Metals. These linkages emphasize the needs of industries in terms of consultancy services and feasibility studies on the commercialization of new technology, the introduction of advanced technology, and quality and productivity improvements (Yoon, 1990). An important issue here is whether the domestic-based research institutes (including the universities) and the banking sector are in a position to develop such a mechanism to enhance the SMIs.

To allow greater public–private sector linkage, an effective policy option would be to encourage the formation of small-scale industry associations within the more important industrial sub-sectors. Through such associations small-scale enterprises could collectively play a more constructive role, including mobilizing funds, encouraging contact with their larger counterparts, and probably establishing joint research efforts and marketing activities. SIRIM, for instance, may be able to create these linkages since it has, in 1986, established a Technology Transfer Centre (TTC) aimed at providing information and services in support of technology transfer. Areas covered include the promotion of the efficient use of technological information, the improvement of the flow of strategic information to local industries, the arrangement of the most equitable terms of technology transfer and the promotion of the effective transfer of technology; the upgrading of the technological capability of

SMIs; the promotion of the commercialization and application of technology; and the enhancement of the diffusion of patent information.

According to Hamzah Kassim (1988a: 171–82), the scope of the TTC covers two principal areas: the setting up of an information-based system that will provide entrepreneurs with the full range of process and product technology suppliers that are of particular interest to the firms in question, and more effective management of the technology transfer process. The latter includes better planning and implementation of the technology transfer programme, covering areas such as technology negotiation, selection, evaluation, and monitoring.

The Consultancy and Technology Transfer Division of SIRIM also provides a special programme for new entrepreneurs to use its high-technology and expensive precision tools and equipment. It also allows them to set up temporary offices at SIRIM for a maximum period of two years. The principal idea is to spawn and develop new and small-scale manufacturing enterprises by helping them to familiarize themselves with the usage of machines and equipment and to introduce them to new techniques at minimum cost. While the objectives set for the SIRIM–TTC are important, it is also of interest to note the functions of the Ministry of International Trade and Industry's Technology Transfer Unit (TTU). These are to ensure that:

1. any agreement entered into between a local manufacturing company (licensee) and its foreign technology licenser will not be prejudicial to the national interest;
2. the agreement will not impose unfair and unjustifiable restrictions or handicaps on the local licensee;
3. the payment of fees, wherever applicable, will be commensurate with the level of technology to be transferred and will not have adverse effects on the country's balance of payments; and
4. a meaningful transfer of technology takes place.

There appears to be some duplication of functions between the two agencies mentioned above, but strictly speaking the TTU of the Ministry of International Trade and Industry is principally responsible for processing and evaluating the terms and conditions of technology transfer entered into by the local licensee and the foreign licenser.

## The S & T Institutional Framework

The expansion of the national S & T system to support industrial development priorities needs a review of the roles and adequacy of the existing R & D support infrastructure in order to facilitate rapid domestic technology development. It is important that the country has a strong and effective co-ordinating ministry or agency through which S & T policies are formulated, implemented, and monitored at the highest level. Although there is such a structure, it does not seem to be effective in providing the right kind of lead in technology development. This issue is particularly critical given the rapid pace of technological advances made in other countries which have a tremendous domestic impact on Malaysia's overall industrial strategy. A strong ministry will ensure that there is an effective mechanism for enhancing S & T management within the country. If the experiences of Japan and South Korea are any guide, efforts must be made to ensure that R & D institutions collaborate closely with other official planning agencies so that long-range plans and budgets can be reviewed on a government-wide basis. This will also ensure that the priority of these R & D structures is to support domestic manufacturers in utilizing advanced technologies (Crawford, 1987a: 67–73).

Appropriately, the above task is to be the responsibility of the MOSTE under which there is already an established framework, including the National Council for Technology Transfer. What is needed is the strengthening of its existing functions and staffing as well as the redefining of some of the organizational or functional roles of the existing S & T framework. The planning and implementation capability within the Ministry must therefore be enhanced, including its expertise in technology assessment and forecasting.

In enhancing the S & T framework, the following criteria have to be considered. First, there must be clear-cut roles and responsibilities in a streamlined S & T administrative system to avoid duplication of activities or functions which will hinder the efficient management of S & T. For instance, it may be useful to identify separate areas of R & D specialization for the public sector and industry to allow for the efficient utilization of the country's manpower and financial resources. In South Korea and Taiwan, for example, telecommunications and the development of very large-scale integrated circuits are two important areas for public sector research activities (Crawford, 1987a: 67–73). Secondly, to ensure an effective system for technological advancement, there must be cohesiveness and consistency in the various phases of policy-making and programme implementation. This also means that there must be full support from industry for all S & T programmes formulated by the Ministry. Such support can be enhanced through private sector representatives at the policy formulation level.

Proposals regarding the reorganization of the S & T institutional framework and the existing research institutes also imply the need for increased budgetary allocations. The basic aim is to increase public sector support for the creation and expansion of national research facilities and capabilities, the benefits of which will be made available to industry. In this respect, the state can provide incentives by instilling competitiveness among the public sector research institutes, which simply means that those which are capable of producing commercially viable research results should be allocated more funds based on their research capabilities (Rahim Bidin et al., 1984: 108). While this will complement the strategy of achieving a higher share of R & D in the country's GNP, it is also in line with the recommendation of the IMP to create a Science and Technology Fund, intended for the financing of long-term research and the creation of R & D infrastructure both of which require substantial initial investments in priority areas.

The quickening pace of industrial development in the future must also be complemented by an increasing public awareness of S & T within the country. To provide the lead, it is important that there is an unequivocal political commitment towards its development. A strong and visible commitment to S & T has proven to be an extremely valuable input, especially in developing countries which have decided to pursue a R & D based industrial development strategy. The effect of this commitment is partly derived from pure leadership characteristics and partly from the signal that identity with S & T gives to others, that a strong and serious commitment exists. An added effect is the associated commitment of resources to R & D programmes.

S & T development is international in scope and practice to the extent that even the most advanced industrial economies need to be well endowed with efficient access mechanisms to maintain an awareness of advances and breakthroughs made in other countries. Some of the techniques are information-database related, some are related to collaborative work which is increasingly international, and some are based on contacts among formal and informal networks of researchers and commercial interests. Such information flows also relate to a country's trade promotion activities, especially with Third World countries. In this respect, the Malaysian government has agreed to establish a South Investment, Trade and Technology Data Centre (SITTDEC) with the aim of promoting trade, technological developments, and investments among the developing countries (Malaysia, 1991b: 225).

Without the benefit of an efficient information gathering and dissemination system on S & T capabilities within the domestic industries and on new technologies, innovations, or products from abroad, domestic technological advances will certainly be handicapped. Neither market mechanisms nor planning will work effectively and both technological forecasting and evaluation will be limited in their usefulness. In comparison to the Asian NICs, not to mention the industrial countries, Malaysia is lagging behind in this respect. This limits its capacity to benefit from new technologies because of the lack of information available to policy-makers, domestic manufacturers, technologists, and academic researchers.

Such a weakness can be attributed to the inadequate systems of locating, acquiring, assessing, and repackaging scientific and technological information and disseminating such information to potential users. The problem can, however, be overcome through a number of measures: collecting and disseminating information in technology registers, data banks, information and documentation centres, and libraries; having mechanisms for access to external data bases and promoting information exchange programmes with overseas research institutes; and improving the information flows to domestic industries, including the small- and medium-scale enterprises.

The S & T information system can be further utilized to build up the database on domestic engineering and technical skills which, in turn, can facilitate S & T human resource development in the relevant agencies or ministries. The database as of 1990 appears to be inadequate for policy analysis of critical manpower needs and for providing information on the precise requirements of industry for specific engineering and technical skills. There is a constant need to review the status of critical skills, including those related to technology development, and this is especially important given the rapid technological changes in all the industrial subsectors. It is also essential that the strengthening of any existing S & T information system covers capabilities for technology assessment and S & T policy analysis.

It is imperative that existing S & T information systems under the various agencies be reorganized/rationalized under one umbrella so that the technology assessment and policy analysis aspects can be directed with uniformity and co-ordination. Under any reorganized framework, the major long-term objectives should include, first, the creation of a conducive climate for the appropriate diffusion of technology (for example, the enhancement of domestic R & D must be regarded as a critical element in the absorption of imported technologies); secondly, the creation of effective linkages between the development of indigenous S & T in Malaysian universities and research agencies and commercial technology needs of the private sector; and thirdly, the purchase of technology (if necessary) for the benefit of local companies, especially the SMIs. The last two aspects would involve co-operation between research agencies and technology users in the private sector.

Information managers, too, must have a conceptual picture of the technology process and an overall understanding of the areas covered. The extent to which they can be drawn into the process depends on the attitude of management and on the corporate culture within an organization. Their important role in overcoming the problems involved in transferring complex technologies must include, among other things, assisting in obtaining complete and reliable information on the technology transfer process, assisting in selecting appropriate technology that is cost-effective, and contributing to the assessment of the effects of policies and government practices and cultural differences on technology transfer (Farkas-Conn, 1988: 47–56). While the state should be responsible for ensuring that the planning is implemented at all levels, the role of industry should be incorporated early in the planning process. Political stability and an environment of continuity and consistency in planning are crucial in order to encourage the participation of the private sector in collaborative S & T projects. Since decision-making often takes place at upper management levels, a scientific awareness must be introduced at these levels. Implementation is the key factor in R & D plans, and it requires a close relationship between public institutions and industry (Curien, 1989: 235–9).

Policies regulating the terms of technology transfer must also be complemented with policies which effectively promote domestic technological capability. Apart from promotional measures (including subsidies and tax incentives, manpower development, etc.), there must also be a selective policy towards foreign technology imports. This selective policy means that the import of foreign technologies can be discriminated against, especially in areas where domestic technology or potential exists. Measures under this policy should include the encouragement of technology imports which have the potential to enhance adaptation and innovative capacities within domestic firms or which can complement domestic capabilities, and the restriction of technology imports which compete with domestic capabilities.

In the long run, a more systematic and comprehensive approach to the issue of technology transfer acceleration is clearly needed. The knowledge acquired through a technology transfer mechanism helps to generate indigenous capacity to adopt and shape the path of technical change, at least in the short run. However, there is a need to find mechanisms that will ensure effective transfer of 'know-how' and 'know-why' among domestic industries in the long term. These mechanisms must be matched with local firms' capacity to absorb, adapt, upgrade, and innovate the acquired technology from abroad. Only then will Malaysians witness an industrial sector that will move to a higher level of technology development and growth.

# Bibliography


Ahmad Tajuddin Ali (1989), 'Direction of Technology Growth in Malaysia', Paper presented at Seminar on Financing Strategies for Technology Growth, Kuala Lumpur, 4 September.

Ahmad Zaharudin Idrus (1988), 'Utilization, Assimilation and Dissemination of Research Results in Malaysia', in United Nations Conference on Trade and Development, *Technology Policies for Development and Selected Issues for Action*, New York: United Nations.

Ahn, Choong Yong (1991), 'Technology Transfer and Economic Development: The Case of South Korea', in Karen Minden (ed.), *Pacific Cooperation in Science and Technology*, Honolulu: East–West Center.

Anuwar Ali (1984), 'The Need for Technical Skills and Innovative Capacity: The Case of Manufacturing Industries in Malaysia', *Akademika*, No. 25, July.

____ (1988), 'Economic Development Strategies in Malaysia: Evaluations of the Past and Challenges Ahead', Paper presented at the First Conference on Asia–Pacific Relations, Foundation for Advanced Information and Research (FAIR), Tokyo, 20–22 April.

____ (1989a), 'Study on the Manufacturing Sector', Report for the HRD Project, Economic Planning Unit, Prime Minister's Department/ILO Project, Kuala Lumpur.

____ (1989b), 'Development Policies for Heavy Industries', in Suh Jang-Won (ed.), *Strategies for Industrial Development: Concept and Policy Issues*, Kuala Lumpur: Asian and Pacific Development Centre and Korea Development Institute.

____ (1991), 'In-Service Training: The Malaysian Experience and Policy Implications', in Karen Minden (ed.), *Pacific Cooperation in Science and Technology*, Honolulu: East–West Center.

Anuwar Ali and Muhd Anuar Adnan (1990), 'Technological Acquisition and Absorption Via Multinational Companies: The Malaysian Experience', *Jurnal Fakulti Ekonomi*, Vols. 21 and 22, Universiti Kebangsaan Malaysia, Bangi, Malaysia.

Anuwar Ali and Zawawi Ismail (1990), 'Global Economic and Technology Trends: Implications for Malaysia and the Universities', Paper presented at the International Conference on University–Industry Interaction, Kuala Lumpur, 21–23 August.

Anuwar Ali, Toh Kin Woon, and Zulkifly Hj. Mustapha (1979), 'Unemployment, Educational Planning and Strategies for Employment in Malaysia', in Cheong Kee Cheok, Khoo Siew Mun, and R. Thillainathan (eds.), *Malaysia: Some Contemporary Issues in Socioeconomic Development*, Kuala Lumpur: Persatuan Ekonomi Malaysia.

Arudsothy, P. (1977), 'Poverty in the Wage Sector and the Problems of Formulating a Minimum Wage Policy', in B. A. R. Mokhzani and S. M. Khoo (eds.), *Some Case Studies on Poverty in Malaysia*, Kuala Lumpur: Malaysian Economic Association.

Asher, S. M. and Inoue, K. (1985), 'Industrial Manpower Development in Japan', *Finance and Development*, 22(3).

Asian and Pacific Centre for Transfer of Technology (1986), *Technology Policies and Planning: Malaysia*, Bangalore, India.

Beaumont, P. G. (1989), 'The Role of Foreign Investment in Industrial Development', Paper presented at the National Seminar on Development towards an Industrialised Economy, Kuala Lumpur, 13–14 November.

Blumenthal, T. and Lee, C. H. (1985), 'Development Strategies of Japan and the Republic of Korea: A Comparative Study', *The Developing Economies*, 23(3).

Brown, T. L. (1985), 'University–Industry Relations: Is There a Conflict?', *Journal of the Society of Research Administrators*, 17(2).

Burgelman, R. A. and Maidique, M. A. (1988), *Strategic Management of Technology and Innovation*, Homewood, Illinois: R. D. Irwin, Inc.

Chacko, G. K. (1986), 'International Technology Transfer for Improved Production Functions', *Engineering Costs and Production Economics*, 10(3).

Chan Yuen Hung (1990), *Industrialisation and Technology Capability Development: A Critical Review of Industrialisation and Industrial Technology Capability Development in Malaysia*, M.Sc. dissertation, University of Sussex.

Chaudhuri, S. (1986), 'Technological Innovation in a Research Laboratory in India: A Case Study', *Research Policy*, 15(2).

Chee Peng Lim (1990), 'International Capital Flows and Economic Development in the Asia–Pacific Region', Paper presented at the International Conference on Financial Institutions and Investment Opportunities in the Asia–Pacific Region in the 1990s, Kuala Lumpur, 11–12 May.

Chen, E. K. Y. and Wong, Teresa (1989), 'The Future Direction of Industrial Development in the Asian Newly Industrialised Economies', in Suh Jang-Won (ed.), *Strategies for Industrial Development: Concept and Policy Issues*, Kuala Lumpur: Asian and Pacific Development Centre and Korea Development Institute.

Chng Meng Kng, Low, Linda, Tay Boon Nga, and Amina Tyabji (1986), *Technology and Skills in Singapore*, Singapore: Institute of Southeast Asian Studies.

Choi Hyung-Sup (1988), 'Technology Development and Industrialisation: Korean Approaches', Paper presented at Workshop on Technology Development for Economic Growth, Kuala Lumpur, 11–13 March.

Christensen, C. H. and da Rocha, A. (1988), 'Perceptions of Brazilian and Foreign Technology', *European Journal of Marketing*, 22(1).

Clifford, M. (1988), 'Stake in the Future: South Korea's Venture Capital Firms Blossom', *Far Eastern Economic Review*, 139(13).

____ (1991), 'South Korea's Electronics Makers Are Slowly Catching Up', *Far Eastern Economic Review*, 154(44).

Commonwealth Secretariat (1985), *Technological Change: Enhancing the Benefits*, Report of a Commonwealth Working Group, Vol. I, London.

Copeland, L. (1986), 'Skills Transfer and Training Overseas', *Personnel Administrator*, 31(6).

Crawford, M. H. (1987a), 'Technology Transfer and the Computerization of South Korea and Taiwan—Part I: Developments in the Private Sector', *Information Age*, 9(1).

____ (1987b), 'Technology Transfer and Computerization of South Korea and Taiwan—Part II: The Public Sector and High-Technology Policy', *Information Age*, 9(2).

Curien, H. (1989), 'Actions to Facilitate Cooperation between Industries, Universities and Other Research Organisations: Attitudes and Experience of Governmental Institutions', *Technovation*, 9(2, 3).

Cutler, R. S. (1989), 'A Comparison of Japanese and U.S. High-Technology Transfer Practices', *IEEE Transactions on Engineering Management*, 36(1).

Dahlman, C. J. and Ross-Larson, B. (1987), 'Managing Technological Development Lessons from the Newly Industrializing Countries', *World Development*, 15(6).

Dahlman, C. J. and Westphal, L. E. (1981), 'The Meaning of Technological Mastery in Relation to Transfer of Technology', *Annals of the American Academy of Political and Social Science*, Vol. 458, November.

Debresson, C. (1989), 'Breeding Innovation Clusters: A Source of Dynamic Development', *World Development*, 17(1).

Djeflat, A. (1988), 'The Management of Technology Transfer: Views and Experiences of Developing Countries', *International Journal of Technology Management*, 3(1, 2).

*Economist* (1991), 'Europe's Two Trade Areas Have Formed an Alliance: There Will Be Gains Aplenty', 321(7730), 26 October–1 November.

Farkas-Conn, I. S. (1988), 'Human Aspects of Information Management for Technology Transfer', *Information Management Review*, 4(2).

Fisher, W. A. (1979), 'Institutional Development of Appropriate Industrial Technology in Developing Countries: R & D and Programmes', in UNIDO Monographs on Appropriate Industrial Technology, No. 1, *Conceptual and Policy Framework for Appropriate Industrial Technology*, New York: United Nations.

Fong Chan Onn (1990), 'Technology Development: The Diffusion of Microelectronic Industrial Machinery in Malaysia', *Jurnal Fakulti Ekonomi*, Vols. 21 and 22, Universiti Kebangsaan Malaysia, Bangi, Malaysia.

Foster, D. (1980), *Innovation and Employment*, Oxford: Pergamon Press.

Fransman, M. (1986), *Technology and Economic Development*, Brighton: Wheatsheaf Books.

Girvan, N. P. and Marcelle, G. (1990), 'Overcoming Technological Dependency: The Case of Electric Arc (Jamaica) Ltd; A Small Firm in a Small Developing Country', *World Development*, 18(1).

Griffin, K. (1969), *Underdevelopment in Spanish America: An Interpretation*, London: Allen & Unwin.

Hakam, A. N. and Chang, Zeph-Yun (1988), 'Patterns of Technology Transfer in Singapore: The Case of the Electronics and Computer Industry', *International Journal of Technology Management*, 3(1, 2).

Hamzah Kassim (1988a), 'Technology Transfer System in Malaysia', in United Nations Conference on Trade and Development, *Technology Policies for Development and Selected Issues for Action*, New York: United Nations.

____ (1988b), 'Technology Development and the Strategy of the Firm', Paper presented at the Workshop on Technology Development for Economic Growth, Kuala Lumpur, 11–13 March.

Hayashi, T. (1983), 'Japanese Educational and Technological Infrastructure: Role of Vocational and Technical Education in Japanese Development', Paper presented at Seminar on the Japanese Experience: Lessons for Malaysia, Penang, January.

Hirono, R. (1986), 'Japanese Experiences in Technology Transfer', in C. Y. Ng,

R. Hirono, and R. Y. Siy, Jr. (eds.), *Technology and Skills in ASEAN: An Overview*, Singapore: Institute of Southeast Asian Studies.
Jaafar Ahmad (1989), 'Investment Priorities and Government Assistance Policies', Paper presented at the National Seminar on Development towards an Industrialised Economy, Kuala Lumpur, 13–14 November.
Jesudason, James V. (1989), *Ethnicity and the Economy: The State, Chinese Business, and Multinationals in Malaysia*, Singapore: Oxford University Press.
Kang Han-Chol (1986), *Promotion of Technology Utilization in Selected Asian Countries: A Report*, Bangalore: Asian and Pacific Centre for Transfer of Technology.
Khor Kok Peng (1983), *The Malaysian Economy: Structures and Dependence*, Kuala Lumpur: Marican and Sons.
Kirkpatrick, C. H., Lee, N., and Nixson, F. I. (1984), *Industrial Structure and Policy in Less Developed Countries*, London: Allen & Unwin.
Koo, B. H. and Kwack, T. (1988), 'Korea's Economic Development Strategy', Paper presented at the First Conference on Asia–Pacific Relations, Foundation for Advanced Information and Research (FAIR), Tokyo, 20–22 April.
Kuhlman, J. A. (1986), 'Industry, Universities, and the Technological Imperative', *Business and Economic Review*, 32(4).
Lee Kye-Woo (1987), 'Human Resources Planning with Special Reference to Technical Education and Vocational Training', in Il Sa-Kong (ed.), *Human Resources and Social Development Issues*, Seoul: Korean Development Institute.
Lee, S. K. (1989), 'Framework for Promoting Industrial R & D in Singapore', Paper presented at Seminar on Financing Strategies for Technology Growth, Kuala Lumpur, 4 September.
Leong, George (1989), 'New Initiatives towards Achieving an Industrialised Economy', Paper presented at the National Seminar on Development towards an Industrialised Economy, Kuala Lumpur, 13–14 November.
Liang, Kuo-Shu and Liang, Ching-ing Hou (1988), 'Development Policy Formation and Future Policy Priorities in the Republic of China', *Economic Development and Cultural Change* (Supplement), 36(3).
Lim, D. (1975), 'Industrialisation and Unemployment in West Malaysia', in D. Lim (ed.), *Readings on Malaysian Economic Development*, Kuala Lumpur: Oxford University Press.
Lindsey, Q. W. (1985), 'Industry–University Research Cooperation: The State Government Role', *Journal of the Society of Research Administrators*, 17(2).
Lo, Fu-Chen and Song, Byung-Nak (1987), 'Industrial Restructuring of the East and Southeast Asian (ESEA) Economies', in Fu-Chen Lo (ed.), *Asian and Pacific Economy towards the Year 2000*, Kuala Lumpur: Asian and Pacific Development Centre.
Lo, Fu-Chen, Kamal Salih, and Nakamura, Y. (1987), 'Asian–Pacific Economy towards the Year 2000: An Overview of Trends and Key Issues', in Fu-Chen Lo (ed.), *Asian and Pacific Economy towards the Year 2000*, Kuala Lumpur: Asian and Pacific Development Centre.
Lo Sum Yee (1972), *The Development Performance of West Malaysia 1955–67*, Kuala Lumpur: Heinemann Educational Books.
Low, Paul (1991), 'Enhancing Export Capabilities of the Malaysian Manufacturing Sector', in Tan Siew Hoey (ed.), *Enhancing Malaysia's Export Capabilities*, Kuala Lumpur: Institute of Strategic and International Studies.
Low Peng Lum (1990), 'Foreign Investment in Malaysia', Paper presented at Seminar on the Inflow of Foreign Investments and their Impact on Small- and

Medium-Scale Enterprises in Malaysia, Kuala Lumpur, 24 February.

Malaysia (1965), *First Malaysia Plan 1966–70*, Kuala Lumpur: Government Printer.

____ (1971), *Second Malaysia Plan 1971–75*, Kuala Lumpur: Government Printer.

____ (1981a), *Fourth Malaysia Plan 1981–85*, Kuala Lumpur: National Printing Department.

____ (1981b), *Labour and Manpower Report 1980*, Ministry of Labour and Manpower, Kuala Lumpur.

____ (1984), *Mid-Term Review of the Fourth Malaysia Plan 1981–85*, Kuala Lumpur: National Printing Department.

____ (1986a), *Fifth Malaysia Plan 1986–90*, Kuala Lumpur: National Printing Department.

____ (1986b), *The National Science and Technology Policy*, Ministry of Science, Technology and the Environment, Kuala Lumpur.

____ (1987), *Annual Report 1986/87*, Industrial Master Plan Sectoral Task Forces, Malaysian Industrial Development Authority (MIDA), Kuala Lumpur.

____ (1988a), *Industrial Surveys 1986*, Department of Statistics, Kuala Lumpur.

____ (1988b), *Labour and Manpower Report, 1985/86*, Ministry of Labour and Manpower, Kuala Lumpur.

____ (1989a), *Mid-Term Review of the Fifth Malaysia Plan, 1986–90*, Kuala Lumpur: National Printing Department.

____ (1989b), *Economic Report 1989–90*, Ministry of Finance, Kuala Lumpur.

____ (1989c), *Industrial Trends Survey, No. 50*, Malaysian Industrial Development Authority (MIDA), Kuala Lumpur.

____ (1990a), *Annual Report 1989*, Bank Negara Malaysia, Kuala Lumpur.

____ (1990b), *Industrial Technology Development: A National Plan of Action*, Ministry of Science, Technology and the Environment, Kuala Lumpur.

____ (1991a), *The Second Outline Perspective Plan 1991–2000*, Kuala Lumpur: National Printing Department.

____ (1991b), *Sixth Malaysia Plan 1991–95*, Kuala Lumpur: National Printing Department.

____ (1991c), *Training for Industrial Development: Challenges for the 1990s*, Report of the Cabinet Committee on Training, Economic Planning Unit, and Ministry of Education, Kuala Lumpur.

Martin, W. S. (1988), 'Research and Development in the 1980s: The Need for Industry–University Cooperation', *Journal of the Society of Research Administrators*, 20(1).

Marton, K. (1986), 'Technology Transfer to Developing Countries via Multinationals', *The World Economy*, 9(4).

Mohd Saufi Haji Abdullah (1986), 'The Role of HICOM in Malaysia's Push towards Becoming an Industrialised Nation', *Kajian Ekonomi Malaysia*, 23(1).

Mohd Yusof Ismail (1989), 'Export Competitiveness: Objectives and Performance of the Industrial Master Plan', Paper presented at the National Seminar on Development towards an Industrialised Economy, Kuala Lumpur, 13–14 November.

Mukerjee, D. (1986), *Lessons from Korea's Industrial Experience*, Kuala Lumpur: Institute of Strategic and International Studies.

Myrdal, G. (1968), *Asian Drama: An Inquiry into the Poverty of Nations*, Harmondsworth: Penguin Books.

Nadkarni, K. (1989), 'The Role of the Industrial Credit and Investment

Corporation of India Ltd. in Financing Technology Growth', Paper presented at Seminar on Financing Strategies for Technology Growth, Kuala Lumpur, 4 September.

Narjes, K. H. (1989), 'Policies and Experiences of International Organisations for the Promotion of Enhanced Interaction between Industries, Universities and Other Research Organisations', *Technovation*, 9(2, 3).

*New Straits Times* (1990), 'Malaysia, Second Choice of Japanese', 19 April.

____ (1990), 'Move to Produce Skilled Manpower', 19 April.

____ (1990), 'SIRIM: $50m Fund Will Boost R & D Activities', 27 April.

Ng, Sek-Hong (1987), 'Training Problems and Challenges in a Newly Industrialising Economy: The Case of Hong Kong', *International Labour Review*, 126(4).

Nurul Islam (1988), 'Economic Interdependence between Rich and Poor Nations', in Noordin Sopiee, B. A. Hamzah, and Leong Choon Heng (eds.), *Crisis and Response: The Challenge to South–South Economic Cooperation*, Kuala Lumpur: Institute of Strategic and International Studies.

OECD (Organization for Economic Co-operation and Development) (1985), *Science and Technology Policy Outlook 1985*, Paris.

Omer, A. (1988), 'Channels and Mechanisms for Transfer of Technology', in United Nations Conference on Trade and Development, *Technology Policies for Development and Selected Issues for Action*, New York: United Nations.

Ozaki, Robert (1991), *Human Capitalism: The Japanese Enterprise System as World Model*, Tokyo: Kodansha International Limited.

Ozawa, T. (1982), *The Role of Transnational Corporations in the Economic Development of the ESCAP Region: Some Available Evidence from Recent Experience*, ESCAP/UNCTC Publication Series B, No. 2, Bangkok.

Park, Eul-Yong (1989), 'Recent Structural Changes in Korean Economy and Foreign Investment', Paper presented at the Asian Economic Development Symposium, Tokyo, 13–14 November.

Park, Yung-Chul (1989), 'Macroeconomic Developments and Prospects in East Asia', Paper presented at the 18th Pacific Trade and Development Conference on Macroeconomic Management in the Pacific: Growth and External Stability, Kuala Lumpur, 11–14 December.

Parry, T. G. (1988), 'The Multinational Enterprise and Restrictive Conditions in International Technology Transfer: Some New Australian Evidence', *Journal of Industrial Economics*, 36(3).

Patvardhan, V. S. (1984), 'The Mechanism of Transfer of Technology and Problems of Adoption: Some Indian Case Studies', in M. J. Campbell (ed.), *Technology Transfer and Social Transformation*, Kuala Lumpur: Association of Development Research and Training Institutes of Asia and the Pacific.

Powell, D. (1989), 'Consortia Just One Answer to High-Tech Woes', *Computing Canada*, 15(12).

Rahim Bidin, Pereira, R. D., and Tan Choon Kok (1984), *Laporan Dasar-Dasar dan Rancangan Pemindahan Teknologi*, Shah Alam, Selangor: Standards and Industrial Research Institute of Malaysia.

Reynolds, L. G. (1964), *Labour Economics and Labour Relations*, Englewood Cliffs; New Jersey: Prentice-Hall.

Rothwell, R. and Zegveld, W. (1981), *Industrial Innovation and Public Policy: Preparing for the 1980s and 1990s*, London: Frances Pinter (Publishers).

Sadasivan, N. (1986), 'The Industrial Master Plan: Why It Has Been Prepared and What Are Its Objectives', Paper presented at the Executive Seminar on the Industrial Master Plan, Kuala Lumpur, 18–19 March.

Sakaiya, Taichi (1991), *The Knowledge–Value Revolution*, Tokyo: Kodansha International Limited.

Saxena, A. N. (1990), 'Strategic Human Resource Development: Role of Technology', Paper presented at the Seventh World Productivity Congress, Kuala Lumpur, 17–22 November.

Seaward, N. (1988), 'Triggering Trade Hopes: Anglo-Malaysian Arms Deal Promises to Boost Business Links', *Far Eastern Economic Review*, 142(41).

Segal, A. (1987), 'Learning by Doing', in A. Segal (ed.), *Learning by Doing: Science and Technology in the Developing World*, Boulder: Westview Press.

Semudram, M. and Mokhtar Tamin (1989), 'Human Resource Development in Malaysia', Paper presented at the Persidangan Ke IV Persatuan Sains Sosial Malaysia, Dasar Ekonomi Baru dan Masa Depannya, Kuala Lumpur, 24–26 July.

Shahid Alam, M. (1989), 'The South Korean "Miracle": Examining the Mix of Government and Markets', *The Journal of Developing Areas*, 23(2).

Shapiro, D. (1986), 'The Unique Alliance Coaxing Taiwan's Biotech Industry into Existence', *International Management*, 41(6).

Sharif, M. N. (1986), *Technology Policy Formulation and Planning: A Reference Manual*, Bangalore: Asian and Pacific Centre for Transfer of Technology.

Siemsen, P. D. (1988), 'Technology Transfer and Licensing in Brazil', *International Journal of Technology Management*, 3(1, 2).

Smith, T. E. (1952), *Population Growth in Malaya*, London: Royal Institute of International Affairs.

Stewart, F. (1978), *Technology and Underdevelopment*, London: Macmillan.

Succar, P. (1987), 'International Technology Transfer: A Model of Endogenous Technological Assimilation', *Journal of Development Economics*, 26(2).

Suh Jang-Won (1989), 'Growth and Industrial Development in the Asia Pacific', in Suh Jang-Won (ed.), *Strategies for Industrial Development: Concept and Policy Issues*, Kuala Lumpur: Asian and Pacific Development Centre and Korea Development Institute.

Sutcliffe, R. B. (1971), *Industry and Underdevelopment*, London: Addison-Wesley.

Takenaka, H. (1989), 'The Japanese Economy and the Economic Development of the Pacific Region', Paper presented at the 18th Pacific Trade and Development Conference on Macroeconomic Management in the Pacific: Growth and External Stability, Kuala Lumpur, 11–14 December.

Thirlwall, A. P. (1989), *Growth and Development: With Special Reference to Developing Economies*, London: Macmillan.

Tuma, E. H. (1987), 'Technology Transfer and Economic Development: Lessons of History', *Journal of Developing Areas*, 21(4).

Tyson, L. (1986), 'Continuing Education in Europe: Cooperation on the Rise', *Training and Development Journal*, 40(11).

United Nations (1983), *Transnational Corporations in World Development: Third Survey*, ST/CTC/46, New York.

____ (1984), *Costs and Conditions of Technology Transfer through Transnational Corporations*, ESCAP/UNCTC Publication Series B, No. 3, Bangkok.

____ (1985), *Transnational Corporations from Developing Asian Economies*, ESCAP/UNCTC Publication Series B, No. 3, Bangkok.

UNCTAD (United Nations Conference on Trade and Development) (1988), *Recent Trends in International Technology Flows and Their Implications for Development*, UNCTAD Secretariat, Geneva.

UNCTC (United Nations Centre on Transnational Corporations) (1988), *Joint*

*Ventures as a Form of International Economic Co-operation*, New York: United Nations.

UNDP/UNIDO (United Nations Development Programme/United Nations Industrial Development Organization) (1985), 'Medium and Long Term Industrial Master Plan Malaysia 1986–95: Overview of the Industrial Master Plan', Vol. 1-1-1 (unpublished report).

UN ESCAP (United Nations Economic and Social Commission for Asia and the Pacific) (1988), *An Overview of the Framework for Technology for Development*, Bangalore: Asia and Pacific Centre for Transfer of Technology.

Urata, S. (1989), 'Recent Economic Developments in the Pacific Region and Changing Role of Japan in the Regional Interdependence', Paper presented at the Second Conference on Asia Pacific Relations, Fukuoka City, Japan.

Wigand, R. T. and Frankwick, G. L. (1989), 'Inter-Organisational Communication and Technology Transfer: Industry–Government–University Linkages', *International Journal of Technology Management*, 4(1).

Wong Poh Kam (1989), 'Critical Skills for Industrial Development', Paper presented at the National Seminar on Development towards an Industrialised Economy, Kuala Lumpur, 13–14 November.

World Bank (1979), *World Development Report*, New York: Oxford University Press.

____ (1989), *World Development Report*, New York: Oxford University Press.

Yankey, G. Sipa-Adjah (1987), *International Patents and Technology Transfer to Less Developed Countries: The Case of Ghana and Nigeria*, Aldershot: Gower Publishing.

Yoo Seong-min (1989), 'Investment Priorities and Government Support Policies in Korea's Economic Growth', in Suh Jang-Won (ed.), *Strategies for Industrial Development: Concept and Policy Issues*, Kuala Lumpur: Asian and Pacific Development Centre and Korea Development Institute.

Yoon, Heon-Deok (1990), 'The Promotional Policy of Small and Medium Scale Enterprises: Korean Experience', Paper presented at the Regional Workshop on the Promotion of Small and Medium Enterprises: Policy Environment and Institutional Framework, Kuala Lumpur, 12–13 November.

Yoshihara, Kunio (1988), *The Rise of Ersatz Capitalism in South-East Asia*, Singapore: Oxford University Press.

Young, K., Bussink, W. C. F., and Hassan, P. (1980), *Malaysia: Growth and Equity in a Multiracial Society*, Baltimore: The Johns Hopkins University Press.

Yu Seong-jae (1989), 'Moving Up the Comparative Advantage Ladder: The Case of Korea's Electronics Industry', in Suh Jang-Won (ed.), *Strategies for Industrial Development: Concept and Policy Issues*, Kuala Lumpur: Asian and Pacific Development Centre and Korea Development Institute.

Zainal Abidin Abdul Rashid (1990), 'Strategies for Human Resource Development', *Jurnal Fakulti Ekonomi*, Vols. 21 and 22, Universiti Kebangsaan Malaysia, Bangi, Selangor.

# Index